The Packet Radio Handbook 2nd Edition

The Packet Radio Handbook
2nd Edition

Jonathan L. Mayo, KR3T

TAB BOOKS Inc.
Blue Ridge Summit, PA

AX.25—pending service mark of ARRL
AMSAT—trademark of The Radio Amateur Satellite Corporation
CP/M—trademark of Digital Research Corporation
HAYES—trademark of Hayes Corporation
IBM—trademark of International Business Machines Corporation
MS-DOS—trademark of Microsoft Corporation
NOVRAM—trademark of Xicor
ProDOS—trademark of Apple Computer Corporation
Teletype—trademark of The Teletype Corporation
TRSDOS and TRS-80—trademarks of Tandy Corporation
UNIX—trademark of Bell Labs
Xerox 820—trademark of Xerox Corporation
Z-80—trademark of Zilog Corporation

SECOND EDITION
FIRST PRINTING

Copyright © 1989 by TAB BOOKS Inc.
First edition copyright © 1987 by TAB BOOKS Inc.
Printed in the United States of America

Reproduction or publication of the content in any manner, without express permission of the publisher, is prohibited. The publisher takes no responsibility for the use of any of the materials or methods described in this book, or for the products thereof.

Library of Congress Cataloging in Publication Data

Mayo, Jonathan L.
    The packet radio handbook / by Jonathan L. Mayo.—2nd ed.
      p. cm.
    Bibliography: p.
    Includes Index.
ISBN 0-8306-3222-0 (pbk.)
    1. Radio—Packet transmission—Amateurs'
manuals. I. Title.
TK9956.M365 1989
621.3841'66   89-4324
                                                        CIP

TAB BOOKS Inc. offers software for sale. For information and a catalog, please contact TAB Software Department, Blue Ridge Summit, PA 17294-0850.

Questions regarding the content of this book
should be addressed to:

    Reader Inquiry Branch
    TAB BOOKS Inc.
    Blue Ridge Summit, PA 17294-0214

Edited by Heilan Yvette Grimes

Contents

Acknowledgments	vii
Introduction	viii
1 What Is Packet Radio?	**1**

Packet Operating Examples • Introducing Packet Radio • Digital Communication Basics • Binary Codes • Packet vs. Other Digital Modes • Packet • Packet Station • Integrated Circuits • Microcomputer Systems • Conclusion

2 The History of Amateur Packet Radio	**20**

The First Packets • Packet Radio • Amateur Packet Radio • Conclusion

3 Amateur Packet Radio Hardware Systems	**34**

TNC • TNC System Trace • Modem • Baud Rates • Amateur Packet Radio Modulation Schemes • Radios • Conclusion

4 Networking and Protocols	**50**

Network Basics • Multiplexing • FDM • TDM • SDM • Amateur Packets (Seep 56) Radio Multiplexing Model • Protocol Basics • OSI/RM • Amateur Packet Radio Protocols • HDLC • AX.25 • V-1 • V-2 • Network Layer • Conclusion

5 Setting Up an Amateur Packet Radio Station	**82**

Selecting a Terminal • Selecting a TNC • Selecting a Radio • Sample Stations • Radio Frequency Interference • Antenna Systems • Conclusion

6 Operating Amateur Packet Radio   **105**
The User Interface • Initial Parameters • Monitoring • BBS Operating • Operating Tips • Conclusion

7 Equipment and Accessories   **142**
AEA • Bill Ashby and Son • DRSI • GLB • HAL • Hamilton Area Packet Network • Heathkit • Kantronics • MFJ • Microlog • Pac-Comm • Richcraft Engineering • Software 2000 • S. Fine Software • TAPR • TNC-2 • NNC • VADCG • WA8DED • Conclusion

8 The Future of Amateur Packet Radio   **165**
High-speed Modems • Digital Radios • SAREX 2 • Software and Protocols • Networking • Packet Contesting • Conclusion

A ASCII Chart   **175**

B The RS-232 Standard   **177**
Introduction • Signals • Signal Levels • Cable Configurations • Null Modem • Flow Control • Nonstandard Implementations • Limitations • Wiring Cables • Conclusion

C Sources   **188**
Packet Organizations • Standards Organizations • Publications • Companies

D Operating Frequencies   **195**
HF • VHF

E An Introduction to Amateur Radio   **197**
Introduction • Licenses • Callsigns • Packet Operation • Conclusion

Glossary   **200**

Bibliography   **216**

About the Author   **220**

Index   **223**

Acknowledgments

The success of *The Packet Radio Handbook*, with its multiple printings and this second edition, has been very gratifying. Packet radio is still a very exciting and dynamic mode, and I hope this book provides you with the information and knowledge you seek. Thank you for purchasing a copy.

Many people and organizations helped me produce this book by providing information. I would like to take this opportunity to thank them and acknowledge their participation: AMSAT, ARRL, Phil Anderson and Lori Elliott of Kantronics, Gil Boelke and Deborah Sanders of GLB, Andrew Demartini of DRSI, Steve Fine of S. Fine Software, Ted Harris, Lyle Johnson of TAPR, Mike Lamb and Mike Forsyth of AEA, Doug Lockhart of VADCG, Hank Oredson, Haroldion Price, Gwyn Reedy of Pac-Comm, Eddie Richey of MFJ, and Wayne Wilson of Heathkit.

I would also like to thank Brint Rutherford, Roland Phelps, and Ray Collins at TAB for their help in making both editions of this book possible.

Introduction

The introduction to the first edition of this book began "Amateur packet radio has been receiving a great deal of attention over the past few years, and justly so." When I originally wrote that sentence, amateur packet radio was expanding rapidly. It was still a relatively new mode, and there were many active amateurs who had not yet heard of packet radio. That has all changed.

Today, there are over 35,000 amateur packet radio stations in operation in North America. The worldwide number has also significantly increased over the last few years. The vast majority of the 30,000 or so packet operators in North America have been operating for several years, predating even the first edition of this book. Now, after spending several years operating packet, these operators have become fairly expert. The remaining packet operators are relative beginners, and they have much more to learn today compared with several years ago.

Packet radio technology has rapidly advanced over the past few years. When the first edition of this book was written, TNC-2s had just entered the mainstream. Today, there are a multitude of TNC-2 compatible enhanced TNCs, multimode digital controllers, high-speed modems, and advanced networking systems. This edition has been thoroughly updated to include information on all the recent advancements in packet radio, and the introductory material has also been revised so that newcomers can easily gain a thorough understanding of packet radio basics.

With this book, I have attempted to bridge the gap between the beginning operator and the more technically astute operator. The first two chapters are introductory and provide the background necessary to understand the concepts presented later in the book. Chapters 3 and 4 are technical and concentrate on the inner workings of packet radio. Chapters 5, 6, and 7 provide information on the operational aspects of packet radio. Chapter 8 provides a summary and look into the future of amateur packet radio. Several appendices are included for reference.

This book is of interest to both beginning and experienced packet operators. I hope you find it enjoyable as well as informative. Should you notice any errors or omissions, please notify me through TAB BOOKS Inc.

This book is dedicated to my parents: Tommie and Ralph.

Chapter 1

What Is Packet Radio?

Tom had been looking forward to this moment for a long time now. Only a few days ago, after reading several articles and talking with a few local operators, he had decided to finally take the plunge. Now that it was Saturday, he was free to drive the hour or so that it took to reach the closest amateur radio dealer.

Sure, he could have had it shipped, but that wouldn't have been the same as bringing it home himself. He rationalized that since he had decided to join the frontier of amateur radio communications, he should at least do it in person. And now it was about to happen.

Tom pulled into his driveway and parked the car. Carefully cradling a package that had been sitting on the passenger seat, he left the car and hurried to the front door of the house. He opened the door and made a beeline to his basement ham shack.

Tom gingerly set the package down on his workbench. Resisting the urge to tear the box open, he fumbled around for a knife and hurriedly slit the tape running along the top. After opening the box and removing the initial layers of packing material, Tom found what he was looking for. Here was the object that he had spent the past three days thinking about: his TNC, his gateway to the world of amateur packet radio.

With this device attached to his microcomputer, Tom knew he would be able to enjoy all the benefits and capabilities of packet radio operation that he had read and heard about. Removing the contents of the box and tossing the instruction manual aside in true amateur fashion, Tom contemplated his next move.

2 What Is Packet Radio?

Tom was perplexed upon examining the TNC. He had never run across anything quite like it in his amateur career. Here was a case about the size of a cigar box with a row of LEDs running across the front and a variety of sockets on the rear panel. Tom finally gave in after examining the unit for a few minutes and was soon busy reading the instruction manual.

About half an hour later, Tom had the unit connected between his microcomputer and two-meter FM rig. He was ready to go. However, Tom was stumped. He did not know what to do next. Tom began to flip through the manual and tried to follow some of the examples, but they did not seem to be working. He turned to another section and began to read, occasionally stopping to type something on the microcomputer's keyboard.

Tom would not emerge from his ham shack for another several hours.

Welcome to the world of amateur packet radio!

PACKET OPERATING EXAMPLES

What follows is just a quick look at the almost endless variety of operating situations that are typical in amateur packet radio.

When Northern California suffered a devastating flood in February 1986, amateur packet radio proved a welcome addition to the usual array of emergency communications provided by volunteer amateur operators. Packet radio was used throughout the flood to provide direct communications with a variety of emergency service organizations, including the California Department of Forestry and the American Red Cross.

Through a network of packet remote bulletin board stations and packet repeaters, emergency traffic could be routed to almost any specific destination in the disaster area. Packet radio offered several important advantages over other systems commonly used in the past. Packet was error free and much faster than ordinary radioteletype. Packet did not require that the messages be recopied by hand when received, nor did the messages have to be retyped when sent to another site. The use of bulletin board stations allowed all sites the flexibility to get their messages at convenient times without having to man the packet station continuously.

Packet radio was also used for emergency communications immediately after the Amtrak train wreck near Baltimore, Maryland, in January 1987. A portable packet station was set up near the wreck site and messages were sent to a local packet bulletin board station. At the bulletin board station, an operator forwarded the messages via both packet and voice.

Most recently, amateur packet radio was put to use to handle emergency traffic during the aftermath of the 1988 earthquake in Armenia. At the request of the Amateur Radio Emergency Communications Organization of the USSR, the ARRL organized the shipment of six portable VHF packet stations to the USSR. The equipment was donated by Tandy, Yaseau, and AEA.

Amateur operators are now beginning to fully realize the potential of packet radio for emergency traffic handling, and the systems in use should grow more efficient as amateurs gain more experience with packet in emergency situations.

Amateur packet radio was also utilized effectively during the 1984 Summer Olympics in Los Angeles. Packet stations were set up to rapidly forward telephone messages across long distances. The system was successfully operated twenty-four hours a day for over eleven days. In the end, over 1300 messages were sent. (See Fig. 1-1.)

Amateur packet radio is used every day by ordinary amateur radio operators the world over, with equally spectacular results. Individuals can transfer messages via packet almost anywhere in the United States through a network of bulletin board stations with an automatic forwarding system. Individuals can use their local bulletin board to send and receive mail, upload and download files (such as newsletters and computer programs), and as a gateway to retransmit their signal on another frequency. Imagine instructing the bulletin board to relay your transmissions to the 20-meter band; an amateur in Pennsylvania running only 5 watts on two meters could then communicate with another amateur on the West Coast or even in a foreign country.(See Fig. 1-2.)

Many packet operators can share a single frequency and select which stations they wish to communicate with. It is not uncommon to have an operator using a bulletin board system, while two other operators are conducting a QSO (communication between two amateur operators), and another two are transferring computer programs—all on the same frequency.

Some amateur packet operators access amateur satellites directly to relay their transmissions with amazing accuracy to specified points around the globe. Amateur packet operators with-

4   What Is Packet Radio?

Fig. 1-1. The packet station used at the 1984 Summer Olympics in Los Angeles. (Photo Courtesy of Jim Koski KT6W, 8-84.)

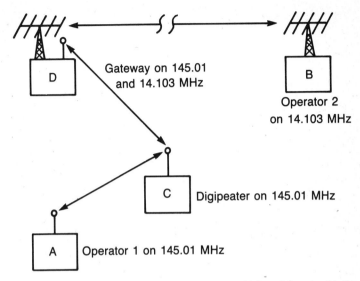

Fig. 1-2. A diagram of a packet radio network in which a digipeater (station C) and HF gateway (station D) are used to allow station A on 2-meters to communicate with station B on 20-meters.

out satellite capability can access the satellites through an intermediary station known as a *teleport*.

You will learn much more about packet radio's capabilities as you progress through this book. Amateur packet radio is a growing, exciting, and largely unexploited (as yet) mode.

INTRODUCING PACKET RADIO

At first glance, packet radio can seem very complicated and confusing. A common assumption is that a packet operator must be an expert in digital electronics and computer programming. While that was probably true a few years ago, packet has advanced to the point where a complete packet station can be set up and effectively operated by almost anyone. And for the microcomputer hobbyist, packet represents an ideal marriage between computers and amateur radio.

But we're getting ahead of ourselves. What exactly is packet radio and what does it have to offer? The purpose of this chapter is to provide an introduction to packet radio's capabilities. This chapter differentiates packet from the other modes of digital communications and provides a solid background so that readers with varying levels of experience can understand the concepts presented in the rest of the book.

Amateur packet radio is a complex digital communications system that uses a high degree of computer technology to obtain a reliable, versatile means of communicating information. Despite the high degree of technology involved in packet radio, it is an easy mode to operate. Packet radio is being routinely used for error-free ragchewing, program and information transfers, and satellite and computer communications.

DIGITAL COMMUNICATION BASICS

In a digital communications system like packet or ordinary *Radio TeleTYpe (RTTY)*, information is transmitted and received in digital form; that is, each piece of information is represented by a digital code. The digital code is made up of one or more elements. These elements can have different states or levels. For example, our decimal number system is a digital code in which each element is one of ten possible levels (0 to 9). The number *one hundred* is represented by the code 100 in our decimal system of ten levels.

In the binary system, there are only two states, represented by a 1 or 0. Thus, there can only be two levels in each element. Each element is called a *BInary digiT (BIT)*. The number *one hundred* (100 decimal) is represented by the binary code 1100100.

The binary system (with two states or levels) is used extensively in digital systems and communications. The two states can be represented in electronic equipment by two different voltages, two different currents, or two different frequencies. In most cases, a multiple state system (three or more levels) is not used because of the increased sensitivity and calibration of the electronics that would be required to differentiate between more than two levels. For example, it is much easier to construct equipment that can differentiate between two voltage levels, often one positive and one negative, than to build equipment that can differentiate between three, four, or more voltage levels.

In order to represent more than two different conditions using the binary system, bits must be combined to increase the number of possible corresponding conditions. For example, one bit is sufficient to indicate if a light is on or off, or if a door is open or closed. (See Fig. 1-3.) However, to represent more complex

Digital Communication Basics 7

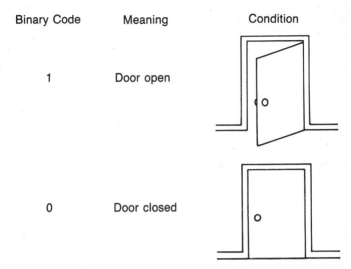

Fig. 1-3. One binary digit can be used to differentiate between two different states such as an open and closed door as shown.

concepts such as the number system or alphabet, several bits are combined. If two bits are used, four different conditions can be indicated. The total number of possible combinations can be found by raising 2 to the power of bits that have been combined (i.e., 2*2=4, 2*2*2=8). Since the alphabet has 26 different characters, 5 bits must be combined for a total of 32 different possible characters. To include the numerals 0 to 9, upper and lower case characters, and some punctuation, a minimum of 7 bits is needed, for a total of 128 possible combinations.

Getting back to the number *one hundred* (100 decimal and 1100100 binary), 7 bits must be combined to represent the number in binary. Six bits are too few (only 64 possible combinations) and 7 bits are too many (128 possible combinations), so we must choose the larger. Although it is relatively easy to convert between binary and decimal, this is not covered in this book. Most computer books contain a conversion table in their appendices should you ever need to switch between binary and decimal.

BINARY CODES

This is how information is transferred using digital communications: Each piece of information is assigned a binary combination,

8 What Is Packet Radio?

which is transmitted to the receiving station(s) where it is reassigned with its original value, provided the sending and receiving station are following the same binary combination assignment (code). This information is usually characters in text; however, it can also be digitized pictures or voice or anything else that can be broken down into discrete pieces of data. (See Fig. 1-4.)

There are several standardized codes in use today for the transfer of text data. Some of them might be familiar to you: *American Standard Code for Information Interchange (ASCII), Expanded Binary Coded Decimal Interchange Code (EBCDIC),* Baudot, and Murray. (An ASCII table is contained in Appendix A.)

Once these binary combinations are generated (whether by a mechanical teleprinter, microcomputer, or some other device), they are often sent to a *MOdulator DEModulator (modem)*. This device generates (modulates) tones (frequencies) which correspond to the state of each bit for transmission over an analog medium, such as a radio link or telephone line. The two tones are given special names. The tone corresponding to the binary 1 is called the *mark*, and the tone corresponding to the binary 0 is called the *space*. (This terminology dates back to the early days of telegraphy when an automatic receiving device would lower a pen on a strip chart when a signal was present making a mark. Of course, when there was no signal, the pen would not touch the strip of paper and a space would result.) These tones are then transmitted by a radio transmitter, or carried via cable, to the receiver(s), where the tones are converted back into digital signals (demodulated) by another modem. This is how binary information gets transferred from sender to receiver.

A transmission medium in which two separate signals may be transmitted in opposite directions at the same time is known as a

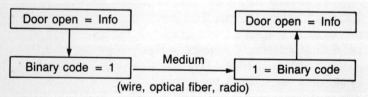

Fig. 1-4. To transfer the condition of the door to another station, the state of the door is noted as information and then the binary representation is sent to the other station. At the receiving station, the binary representation is reassigned the same information.

full duplex channel (i.e., two stations may transmit to each other at the same time). A transmission medium in which two separate signals may be transmitted in opposite directions but only one signal may be transmitted at a time in either direction is known as a *half duplex* channel (i.e., two stations may transmit to each other but not at the same time). A transmission medium in which only one signal may be transmitted at a time in only one direction is known as a *simplex* channel (i.e., one of two stations transmits and the other receives). A half duplex channel can be considered a simplex channel that can reverse direction between stations.

PACKET VS. OTHER DIGITAL MODES

One of the questions most frequently asked about packet radio is how it differs from the other modes of digital communications available to amateur radio operators. *Continuous Wave (CW)*, Baudot RTTY, ASCII RTTY, and *AMateur Teletype Over Radio (AMTOR)* are the other major digital communication systems in use today in amateur radio. This section discusses the capabilities of each of these digital communication systems and then details the capabilities of packet radio. This discussion should give you a good understanding of what packet radio has to offer above and beyond the other modes.

CW is the oldest form of digital communications. It uses an uneven form of coding, usually the Morse code. In an uneven code, the number of elements that make up each character are not equal; thus, some characters have more elements than others. Morse code contains most characters needed for communications and requires very simple equipment for transmission and reception. Since the advent of computerized keyboards and decoders, CW can be sent very quickly. However, the slightest bit of interference or imperfect sending can reduce the decoder's ability to accurately copy the code. Compared to other forms of digital communications, CW leaves a lot to be desired. On the plus side, CW is the only form of digital coding which can be easily copied by ear without the aid of decoding equipment.

Baudot RTTY uses an even form of coding known as the Baudot (or Murray) code. In the Baudot code, each character is made up of five mark and space elements (or bits). However, there are only thirty-two possible combinations using a five level code. Therefore,

Baudot includes two different character sets, figures and letters. The character sets are alternated as needed.

Baudot RTTY operation usually takes place at speeds of 45, 50, or 75 baud using either solid state equipment or mechanical teleprinters. In most cases, the baud rate is roughly equal to the number of elements (or bits) sent per second. A more accurate definition of baud will be given in Chapter 3.

ASCII RTTY was first legalized in 1980 by the FCC for amateur use in the United States in response to the wide proliferation of computer equipment which uses the seven element ASCII code. Its primary advantages over Baudot RTTY are its speed (usually 110 or 300 baud) and its 128 possible characters. Solid state equipment is usually used for ASCII operation, but mechanical teleprinters can also be used. In operation, ASCII RTTY is very similar to Baudot RTTY except for the coding used.

AMTOR introduced two new dimensions to amateur RTTY communications; error-checking and time diversity. These two topics are covered in greater detail in a later chapter. AMTOR was first legalized for amateur use in 1983. AMTOR uses a special even coding in which there is a constant ratio of mark and space elements. If the received character does not have the proper ratio, it is presumed erroneous. AMTOR operates at 100 baud and because of error-checking is much more reliable than standard Baudot or ASCII RTTY.

PACKET

Finally, we get to packet. Packet is the most advanced form of digital communications available to radio amateurs. The primary advantages of packet radio are speed, networking, error-checking, and efficient use of frequency space.

Packet radio operates using a standard digital communications networking technique known as *Carrier-Sense Multiple Access with Collision Detection* (CSMA/CD). Put simply, this means that a packet station will not transmit when the frequency is busy. It waits until the frequency is clear and then transmits a short burst (frame) of information. Because packet transmissions are very short, many packet stations can be on the same frequency without interfering with each other. A line of text that takes 30 seconds to type can be transmitted in a fraction of a second.

Should two packet stations transmit on the same frequency at the same time, their transmissions may interfere with each other (collide). If a collision should occur, each station will wait a random length of time and try again. Most likely, one station will wait a shorter length of time and thus transmit before the other station. Its carrier prevents the other station from transmitting until it is finished.

Most packet activity today is at 1200 baud on the VHF bands and 300 baud on HF. Soon packet will be operating at 9600 baud and up on VHF as modem technology advances.

Packet's error-checking follows the *High-level Data Link Control (HDLC)* format. The data entered by the user is grouped together in bundles of (usually) 128 characters. The binary digits (bits) which compose the data and any other information to be transmitted (such as the sending and receiving stations' callsigns) are put through an extensive polynomial expression and a number unique to the specific data being transmitted is generated. This number is known as the *Frame Check Sequence (FCS)*. The FCS is sent along with the data. When the receiving station gets the data and the FCS, it recomputes the FCS using the same equation and compares it with the one received with the data.

If the two FCSs match, the data is assumed error-free and an *acknowledgement (ACK)* is sent to the transmitting station. If they do not match, the data was not received exactly as the transmitting station sent it, so the receiving station ignores the transmission. The transmitting station retransmits the data after a period of time.

Of course, for all this to work, the two stations must be using compatible equipment and the same protocols. The equipment in packet radio consists of three main components in addition to a transceiver; a terminal, a *Terminal Node Controller (TNC)*, and a modem. The two modems must observe the same standard; usually Bell 202 on VHF and Bell 103 on HF. The two TNCs may be of different manufacture, but they must use the same protocol.

Do not confuse amateur packet radio TNCs as shown in Fig. 1-5 with RTTY Terminal Units (Fig. 1-6) and modems (Fig. 1-7).

Protocols define the format of the information sent. The protocol organizes the information to be transmitted into *frames*. A protocol also defines what steps are to be taken by the TNC under different circumstances. Networking procedures are another area which the protocol defines.

12 What Is Packet Radio?

Fig. 1-5. Two amateur packet radio TNCs (a Heathkit HD-4040 and a Kantronics KPC-2).

Fig. 1-6. The Kantronics Interface II, a RTTY Terminal Unit.

There is one main amateur packet radio protocol in use today: AX.25. AX.25 was developed by *AM-RAD* (Amateur Radio Research and Development Corp.) and *RATS* (Radio Amateur Telecommunications Society) and is a modified version of the X.25 commercial protocol. The AX.25 protocol was developed in 1982 to provide some additional needed capabilities not present in earlier protocols. AX.25 offers the advanced capabilities needed for an extended packet network. Today, AX.25 is the most widely used protocol. (Protocols are covered in much greater detail in Chapter 4.)

Fig. 1-7. The Flesher TU-1200 modem.

Since the sending and receiving stations' IDs (usually their callsigns) are included in the frame as the address, the frame can be routed through different intermediary stations to reach its destination. These intermediary stations can include digipeaters (simplex packet repeaters), satellites, and HF gateways. There is great potential in using packet's networking capabilities to link large areas of the world together.

It is easy to see how much more versatile packet radio is than the other forms of digital communications. While ordinary RTTY can only be relayed through duplex repeaters, packet allows for multiple controlled relays through simplex repeaters. Packet's controlled transmissions assure error-free reception, and packet's high speed reduces the time a packet station must transmit. Packet radio can be used for simple ragchewing, both local and DX, as can the other modes; however, packet offers so much more. Today, packet radio is as simple to operate as ordinary RTTY. No special programming or computer knowledge is necessary, and you do not even need to have a microcomputer; almost any communications terminal will work.

PACKET STATION

The basic packet radio station is composed of four main components: the terminal/computer, a TNC, modem, and transceiver. (See Fig. 1-8.) Originally, the name *Terminal Node Controller* was derived from its use in packet communications networks. Al-

14 What Is Packet Radio?

Fig. 1-8. The amateur packet station at KR3T.

though TNCs can be found at intermediate nodes in a packet network, the name refers to a controller of end (terminal) nodes in the network. Networking is covered in greater detail in Chapter 4.

The TNC is the heart of a packet radio station. The TNC organizes and controls the transmission/reception of the data. Its functions are discussed in greater detail in Chapter 3. The TNC comes in three different flavors: a software package running on a computer with an external modem, a separate hardware circuit board with external modem, and a separate hardware circuit board with a built-in modem. By far, the hardware board with built-in modem is the most popular. It may be used with a dumb data terminal or a computer running terminal emulation software.

The software based TNC uses the processor in the computer to handle the TNC's various functions. An external modem is then added to the computer, and the computer serves as both a TNC and a terminal. However, the software approach consumes memory and processor time, and because of the complexities involved in writing the software, along with the inability to easily produce it for a wide range of computers, the software approach has not become very popular. The microcomputer-based software approach has many advantages, but at the present time they do not outweigh its disadvantages compared to the hardware-based TNCs. The differences are discussed in greater detail in Chapter 3.

The other components of a packet station are the terminal and the radio transceiver. The choice of a terminal system is discussed in Chapter 5. The transceiver is usually a common 2-meter FM rig; however, packet operation occurs on other frequencies as well. Almost any 2-meter transceiver or handheld will do in the beginning.

Setting up a packet radio station, like most other modes, varies greatly in cost depending on what equipment is needed and what kind is acquired. A complete system starting from scratch could cost from as little as $300 up to over $2000. If you already have a computer or terminal and a transceiver, the cost varies from $50 for a software package and a modem up to $325 for an expensive TNC with a built-in modem. Most TNCs with built-in modems sell for around $150. More information on setting up an amateur packet radio station is given in Chapter 5.

INTEGRATED CIRCUITS

In order to fully understand packet radio, it is helpful to have some understanding of integrated circuits and computer systems. *Integrated circuits (ICs)* have truly revolutionized electronics over the past two decades. Without ICs, it is safe to assume that most of the electronic devices we have today (including TNCs) would not have been developed.

Integrated circuits are easy to recognize. They are usually small rectangular chips made from a black plastic material with metallic leads coming out the sides. But there is much more to an IC than this. Sealed in the plastic housing is a miniature square of silicon material measuring a few millimeters on each edge. Etched on this chip of silicon are the electronic circuits that determine the function of the IC.

Typical IC circuits consist of many transistors and gates. Memory chips often contain hundreds of thousands of transistors. There are six main gates: AND, OR, XOR, NAND, NOR, and Inverter. These gates are designed to carry out the rules of Boolean algebra which is based on the binary system. Each gate gives a different result depending on the combination of bits applied to them. (This book won't go into detail on how each gate specifically operates or how different types of gate circuits are designed.)

The construction of integrated circuits is a processing marvel. Successive layers of resists, dopants, metals, and other materials are applied to a silicon wafer to develop the desired circuits. When all the layers are finished, the wafer is broken into hundreds of chips and tiny wires are attached to the inputs and outputs of the chip.

The chip is then sealed in plastic with the tiny input and output wires attached to the metal leads sticking out of the plastic.

ICs have been designed to handle almost any desired digital task from *central processing units (CPUs)*, to memory, to multiplexers and on and on. Were you to open the case of any computer, you would see rows and rows of ICs covering the circuit boards, each with a specific job to do. However, ICs do not work alone. They must have support components to define and regulate their operation. These support components consist of connecting circuits, resistors, capacitors, crystals, and so forth.

MICROCOMPUTER SYSTEMS

With the advance of the microcomputer revolution, more and more amateur radio operators have one or more micros performing a variety of functions in the ham shack. Whether or not you decide to use a microcomputer as a communications terminal in your packet station, it is helpful to know and understand a little about microcomputer systems because they are in such wide use in packet radio.

Microcomputer systems are made up of two major components: hardware and software. Hardware is what most people visualize with when they think of microcomputers. Some examples of hardware are physical devices such as *Cathode Ray Tubes (CRTs)*, keyboards, disk drives, printers, modems, TNCs, and their associated circuitry (see Figs. 1-9 and 1-10). Before going any further, let's take a look at what makes up a microcomputer.

A basic microcomputer is actually composed of only four different sections: input, CPU, memory, and output. Input can take the form of a keyboard, disk drive, or anything else which allows for the input (entry) of information into the computer. The CPU is the brains of a computer. It manages the flow of information to and from the other components and performs arithmetic operations.

Memory is used for the storage of data and programs. It can be written to and read from (as with RAM, which stands for Random Access Memory); however, some forms of memory cannot be written to once initially programmed (like ROM, Read Only Memory). More information on the different types of memory is given in Chapter 3.

Microcomputer Systems 17

Fig. 1-9. Two 5.25-inch disk drives.

Fig. 1-10. An Epson MX-80 printer.

Output is where the CPU directs any information to be sent to the user or for storage on an external device. Some examples of output are display monitors, disk drives, and printers.

You might have noticed that disk drives are listed as examples of both input and output. This is perfectly alright, as some devices are used for both storage and retrieval. These devices fall under the header of *Input and Output (I/O)*. But all this hardware is useless for its intended function without software. An apt analogy would be a car full of gas and ready to go but without a driver.

Software tells the hardware what to do and how to do it. It is simply a list of instructions telling the computer how to accomplish a task. Software comes in many different forms and on various *media*. Media is the physical device on which the software instructions are stored. Some typical microcomputer storage devices include permanent memory (ROM), cassettes, and most commonly, floppy disks. (See Fig. 1-11.) Software stored on permanent memory chips is usually called *firmware* and can be found built into the computer and in plug-in cartridges. Most often rudimentary programs such as system monitors and basic operating systems are included in permanent memory on the computer's *motherboard*. The motherboard usually contains most of the basic circuitry of the computer.

A system monitor program allows the user to look into memory locations and alter their contents. It is very helpful in diagnosing problems with other programs; however, its use does require some technical knowledge about the architecture of the computer. An *operating system* is an extensive program that tells the computer how to operate. It assigns memory locations for various functions, controls the external storage and retrieval of information, and the other control functions needed for the operation of the computer system. A *disk operating system (DOS)* is used to control access to the disk drives. Some examples of current microcomputer disk operating systems in use today are MS-DOS, TRSDOS, CP/M, UNIX, ProDOS, and OS-9.

Once an operating system has been loaded into the computer's memory, it is possible to load and execute *application programs*. Application programs include word processors, spreadsheets, telecommunications, and of course games. While most operating systems and system monitors are written in *machine code* (the most rudimentary programming language which consists of the actual binary information that is stored in the computer's memory) or assembly language, application programs are written in a variety of languages. Computer languages vary from low-level such as machine and assembly language which deal directly with the computer specifically, to high-level which often have a syntax similar to the

Fig. 1-11. A 5.25-inch floppy disk and a 3.5-inch disk.

English language and thus require much less hardware knowledge. Some examples of high-level languages are *Beginners All-purpose Symbolic Instruction Code (BASIC)*, which is included with nearly all microcomputer systems, *FORmula TRANslator (FORTRAN)*, Pascal, and *COmmon Business Oriented Language (COBOL)*. The merits of programming in one language or another are beyond the scope of this book.

CONCLUSION

This chapter has covered mostly background material which is necessary for an understanding of amateur packet radio. Starting with several operating examples, the chapter progressed to binary coding, digital communications, packet radio, and finally integrated circuits and microcomputer systems.

The next chapter provides a comprehensive look at the history of amateur packet radio.

Chapter 2

The History of Amateur Packet Radio

Many years ago, the words *packet radio* would have brought blank stares from most amateurs; today, packet radio is the center of intense discussion and debate in the amateur community. The history of amateur packet radio is very interesting, and for the newcomer to packet radio, knowing its history will help explain the evolution of packet, why certain operating practices occur, and the root of some of the terminology. Some terms are used in this chapter that have not yet been introduced; they are explained in later chapters. A brief explanation of most terms can be found in the glossary. So without further discussion, let's start at the beginning.

THE FIRST PACKETS

The first studies of packet networking were conducted by the Rand Corporation in 1964. The term *packet* was introduced by D.W. Davies of the British National Physics Laboratory in 1965. Work started on developing an actual packet network for the *United States Advanced Research Projects Agency (DARPA)* in 1969.

The DARPA network, named ARPANET, was set up by Bolt Beranek, and Newman, Inc., and included packet switching. Other packet switching networks developed around the world. Today there are many private and government-owned packet switching networks.

PACKET RADIO

The first packet networks were all cable-based (i.e., transmission was over cables). The first large-scale packet radio experiments, where transmission was over radio frequencies, began in 1970. One of the largest and most significant packet radio networks was ALOHANET based at the University of Hawaii.

ALOHANET linked a number of computers and users together using packet switching technology over radio links. (See Fig. 2-1.) ALOHANET had access to ARPANET and satellite relays. ALOHANET was established to investigate the mathematical and practical aspects of a random access packet system. Many of the techniques used in amateur packet radio systems were first developed as a result of ALOHANET.

ALOHANET operated on two frequencies: 407.350 MHz and 413.475 MHz. 407.350 MHz was used to transmit from user terminals to the central computer (called the *menehune*). 413.475 MHz was used by the menehune for transmissions back to the user terminals. Repeaters were included for terminals located too far from the menehune for direct communications.

Each radio station was called a *node*. The equipment at each node consisted of a standard commercial VHF radio, a 9600 baud

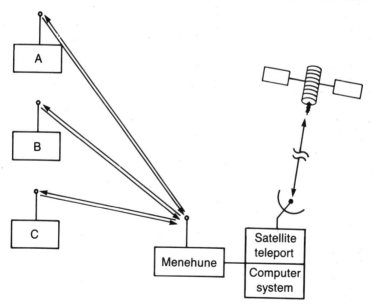

Fig. 2-1. A diagram of the ALOHA packet radio network.

Bits Per Second (BPS), Differential Phase Shift Keying (DPSK), modem, and either a *Terminal Control Unit (TCU)*, or a *Programmable Control Unit (PCU)* for nodes with more than one user terminal attached to it. The TCU and PCU are the equivalent of today's TNCs.

Many different protocols and contention schemes were experimented with on ALOHANET. For more information on ALOHANET, see the following references:

- *Packet Radio* by Rouleau VE2PY and Hodgson VE2BEN from TAB BOOKS (#1345)
- *Computer Networks* by Tanenbaum from Prentice-Hall

AMATEUR PACKET RADIO

Amateur packet radio has an interesting history. The following sections describe the development of amateur packet radio over the years.

Canada: The Initial Sparks

Amateur packet radio operation got its start in Canada in April 1978 when the Department of Communications (DOC) announced that it intended to change the regulations to allow packet radio operation on 220 MHz. Several Canadian amateurs immediately began investigating packet radio operation and were soon busy developing the necessary equipment for packet operation. The first amateur packet radio transmissions occurred on 31 May 1978 in Canada.

The first packets were sent using a packet radio system developed by the Montreal Amateur Radio Club. Their network was named MP-net and was the first amateur packet radio system anywhere in the world. The Montreal group was using a complete homebrew system utilizing a CSMA/CD protocol. The transmission speeds were both 1200 baud and 2400 baud AFSK on the 220 MHz band.

A description of the MP-net is given by two of the participants in their book *Packet Radio* by Rouleau VE2PY and Hodgson VE2BEN. The authors each published an article on the MP-net system. The first, titled "The Packet Radio Revolution," was written by Rouleau and appeared in the December 1978 issue of *73 Magazine*. Hodgson's article, "An Introduction to Packet Radio," appeared in the June 1979 issue of *Ham Radio Magazine*.

Amateur Packet Radio 23

The next Canadian group to get a packet system up and running was the Vancouver Amateur Digital Communication Group (VADCG). VADCG was founded by Doug Lockhart in January 1979. The VADCG system is the best known of the Canadian systems because of their work in spreading packet to the United States.

The first hardware device dedicated to handling packet radio communications was developed by the VADCG. (See Fig. 2-2.) This device was given the name *Terminal Node Controller*, which has stuck with us to the present day. An amateur packet radio protocol was also developed by this group based on *Bit Oriented Protocols (BOP)* such as the HDLC protocol.

The Vancouver system was configured as several users with terminals and TNCs communicating with a central device called a *Station Node Controller (SNC)* at 1200 Baud on 144 MHz. When individual users first connected to the SNC, they were assigned a numeric address. This address was included in all transmissions to that user's station. All communications between users took place through the SNC.

Fig. 2-2. A partially-assembled original VADCG TNC.

The VADCG sold their TNCs to other amateurs in the United States (although packet radio was not yet legal for United States amateurs) and other countries. They published the first amateur packet radio newsletter, *The Packet*. The VADCG also transmitted packet beacon messages on 14.0765 MHz so others outside of Canada where packet operation was not legal could check the operation of their TNCs.

The third Canadian group to get a packet system operational was based in Ottawa. The Ottawa group first began operation in 1980, and their system consisted of a polling protocol and SNC. Transmissions were at 9600 baud using FSK modulation techniques.

United States: The Flames Take Hold

Amateur packet radio operation was not legal for United States operators for about eighteen months after it was legalized in Canada. Finally, on 17 March 1980, the Federal Communications Commission (FCC) authorized ASCII transmissions in response to the wide proliferation of ASCII-based computer systems. The first United States amateur packet radio system run by United States amateur operators was put up near San Francisco, CA. This system was based around the first United States digipeater which was activated on 10 December 1980. This digipeater was developed by Hank Magnuski KA6M and consisted of a homebrew TNC and a modified VADCG protocol.

On the East Coast, the Amateur Radio Research and Development Corporation (AMRAD) began publishing information on packet radio in the AMRAD newsletter. The first amateur packet radio station on the East Coast was activated by Bill Moran W4MIB on 4 May 1981. The East Coast systems were also using a modified VADCG system.

Packet remained in the experimenter's workshop and out of the amateur radio public's eye until the October 1981 issue of *QST Magazine* was published by the ARRL. This issue of *QST* contained an interesting article on packet radio titled "The Making of An Amateur Packet-Radio Network" written by Borden K8MMO and Rinaldo W4RI. The article announced what was to be the first of many ARRL Amateur Radio Computer Networking Conferences on 16 October 1981. This article had a significant impact on amateur packet radio by introducing packet to the general amateur population.

The first ARRL Computer Networking Conference was held on 16 October 1981 in Gaithersburg, MD. It was designed to serve as a general meeting of North American amateurs who were interested in developing packet radio. One of the attendees, Den Connors KD2S, would later do much to add towards that goal.

Tucson: Fire in the Desert

After attending the networking conference, Den Connors returned to Tucson, AZ, and on 29 October 1981 gave a talk at the Tucson chapter of the IEEE computer society. His talk was centered on amateur packet radio; specifically, developing a packet system based on the 6809 microprocessor. Several people in the audience were interested in Den's speech and decided to meet at a later date to discuss amateur packet radio activity in Tucson.

The meeting was held on 6 November 1981 and six people were in attendance (Lyle Johnson WA7GXD, Mark Baker, Dave McClain N7AIG, Marc Chamberlain WA7PXW, Jerry Clark K7KZ, and Den Connors KD2S). The result of this first meeting was the formation of the Tucson Amateur Packet Radio Corporation (TAPR), and a commitment to develop a complete TNC with onboard modem and power supply.

At the time of the first TAPR meeting, the cost involved in equipping an amateur packet radio station was around $300 for the VADCG TNC printed circuit board, parts, a modem, and a power supply. TAPR hoped to be able to develop a TNC with all the necessary components and accessories at half the cost. With this goal in mind, they set off on the development of what would be known as the Alpha TNC.

The Alpha TNC was produced several months later and was based on the 6502 microprocessor. Only twelve Alpha boards were made and distributed to selected participants for testing in April 1982. The Alpha board served as a trial run for TAPR, helping them to learn more about the details involved in producing and distributing TNCs.

In July 1982, TAPR began publication of what has become the most popular amateur packet radio newsletter, the *Packet Status Register (PSR)*. Originally a bi-monthly publication, *PSR* has been changed to a quarterly publication. The original intent of *PSR* was to provide a source of news about amateur packet radio activities and developments in the United States along with technical information for those interested in following TAPR's progress in developing their TNC.

Amateur packet radio newletters and magazines are currently in a state of flux. The monthly *Packet Radio Magazine (PRM)* is now defunct. *PRM* was published by the Florida Amateur Digital Communications Association (FADCA); they have now resumed their own newsletter called the *FADCA > BEACON*. *CTM magazine*, which also published many packet related articles, has also folded. The ARRL's biweekly *Gateway* newsletter serves as a forum for activity-and date-oriented material.

On 9 June 1982, the first packets were transmitted by an Alpha TNC. Later, on 18 June, an Alpha TNC successfully received transmitted packets. With the knowledge gained from the testing of the Alpha TNCs, TAPR began to prepare an updated TNC design for the next stage in their TNC development program, the Beta TNC.

The Spreading Fire

By this time, many amateurs in the United States had begun developing packet radio systems. Several other packet groups had sprung up around the country. Many systems were using the modified VADCG protocol first introduced in California. The TAPR Alpha TNC was using a prototype testing protocol. AMRAD and RATS had developed a protocol based on the CCITT X.25 protocol which they called AX.25. By mid-1982, there was some concern about compatibility between systems, and with the upcoming launch of OSCAR 10 and the prospect of wide area packet communications between the different systems, amateurs wanted to establish a nationwide standard protocol.

In October 1982, the president of AMSAT, Tom Clark W3IWI, called a meeting of United States packet groups with the intention of agreeing on a standard protocol. The three contenders for the standard protocol were VADCG, TAPR/DA, and AX.25. The outcome of the meeting was a new version of AX.25 which supported digipeaters. Thus, AX.25 became the standard protocol for amateur packet operation in the United States.

Tucson: Still Burning

Back in Tucson, work was continuing on the Beta TNC; now to utilize the AX.25 protocol. The Beta TNC was to be distributed to a much larger number of users for testing. Around 170 Beta TNCs were eventually distributed. By December 1982, the Beta TNC was nearing completion. The first few boards had been assembled and operated properly, and the printed circuit boards and components were delivered. The protocol and other software development was being handled by Harold Price NK6K, Dave Henderson KD4NL, and Margaret Morrison KV7D.

At the end of 1982, there were approximately 200 TNCs in existence in North America.

The finished Beta TNC consisted of a 6809 microprocessor, a hardware HDLC system, 24K of ROM, 6K of RAM, and LED status indicators all on a 8 × 8 in. PC Board. The Beta TNCs were sent to numerous sites for testing including St. Louis, Los Angeles, Washington, D.C., and Chicago. In the middle of all the Beta TNC activity, TAPR held their first annual meeting on 5 February 1983. The software underwent many modifications and revisions, and the comments of all Beta TNC testers were noted.

The second ARRL Amateur Radio Computer Networking Conference was held in March 1983 in San Francisco. The published proceedings from this conference contain several papers on such topics as the AX.25 protocol and implementing the functions of a TNC in software.

The July 1983 issue of *Ham Radio Magazine* features the Beta TNC on the cover and includes the first of a two-part article on packet radio titled "Amateur Packet Radio: Part 1" by Morrison KV7D and Morrison KV7B. The August issue contains the second part of the article which covers the technical aspects of the Beta TNC.

Also in July 1983, a meeting was held at the home of Harold Price to discuss the linking of digipeaters to allow for long range communications until true networking could be developed. The end result was WESTNET, a group of digipeaters running up the coast of California. A proposal was made to extend the number of digipeaters allowed for each transmission from three to eight; thus allowing for longer distance connections. On the East Coast, EASTNET and SOUTHNET were set up with the same goals in mind. EASTNET covered the East Coast north of the Carolinas, and SOUTHNET covered south of the Carolinas to Florida.

The Fire Grows Larger

Now that the Beta TNC test period was coming to an end, TAPR was gearing up for the full-scale production of TNC kits for distribution to the mass amateur radio market. Prototype TNC boards were constructed in August 1983, and preparation made for general distribution. On 21 November 1983, the first shipment of 100 TNCs was ready.

The TAPR TNC was supplied in kit form. The kit and assembly manual received rave reviews from the amateur community. The PC board was of excellent quality and the manual was very informative with an easy-to-follow assembly section. All in all, over 2500 TAPR TNC kits were sold by TAPR.

Also in August 1983, another TNC, called the PK-1, was introduced by GLB Electronics. The GLB TNC initially only supported the VADCG protocol; however, AX.25 was added shortly. In the GLB TNC, the HDLC system was handled using software rather than hardware as in the TAPR TNCs. This resulted in a strange situation: a user could not type on the terminal and receive frames over the radio at the same time.

By the end of 1983, there were approximately 650 TNCs in existence in North America.

In February 1984, the first WORLI packet bulletin board system (BBS) was put on the air by Hank Oredson WORLI. This system has become the standard packet BBS in the United States.

The third ARRL Amateur Radio Computer Networking Conference was held in April 1984 in Trenton, NJ. The Dayton Hamvention was held two weeks later and the packet radio forum attracted over 300 attendees. At the 1984 Dayton Hamvention, TAPR's president, Lyle Johnson, was presented with the first Technical Achievement Award for his work on the TAPR TNC. Later that year, in August, Doug Lockhart received the CRRL Certificate of Merit in recognition of his pioneering work in amateur packet radio. Today, Doug Lockhart is widely recognized as "the father of amateur packet radio."

Another significant event occurred at the 1984 Dayton Hamvention; AEA introduced the first TAPR TNC clone. (See Fig. 2-3.) TAPR had made arrangements to sell the rights to their TNC design for a flat fee of five hundred dollars. This step was taken to encourage commercial manufacturers to get involved with packet radio. The AEA PKT-1 is identical to the TAPR TNC in most

Fig. 2-3. The AEA PKT-1.

respects. AEA began to run a series of full page ads in major amateur publications for their TNC.

By the end of 1984, the number of TNCs in existence in North America had grown to approximately 2500.

On 25 March 1985, the FCC eliminated the requirement that a description of all special digital codes used above 50 MHz be included in the station log.

The fourth ARRL Amateur Radio Computer Networking Conference was held on 30 March 1985 in San Francisco.

In March 1985, Heathkit introduced its TAPR TNC clone. Their initial production run of 500 units in April sold out in three weeks. At the 1985 Hamvention, Kantronics introduced their TNC named the "Packet Communicator." In late April, TAPR announced that it was stopping production of its TNC.

TAPR felt that the commercial TNC clones were being produced faster and at a lower cost than TAPR could accomplish. TAPR decided to sell their remaining stock of TNCs at discount prices and clear the way for more developmental work. TAPR's next step was the TNC-2.

The TAPR TNC-2 was a new TNC design based on the Z-80 microprocessor. It contained many new features such as lower power consumption and smaller size than the original TAPR TNC (now called the TNC-1). The TNC-2 development went rather smoothly and test versions were distributed in May 1985.

The July and August issues of *QST Magazine* each contain an article on packet radio written by Harold Price. The August issue featured a TAPR TNC-2 on the cover. (See Fig. 2-4.)

In July 1985, TAPR announced they would begin accepting orders by telephone for the TAPR TNC-2 kit beginning at 9:00 AM on Monday, 19 August 1985. There are only 300 TNCs available and only one kit was allowed per customer—first come first served.

On Monday morning, the telephones at TAPR headquarters began to ring, and they did not stop. Orders for the TNC-2 were coming in at a phenomenal rate. Around noon, TAPR received a call from one of the telephone company engineers; he told them that the level of incoming calls had completely saturated the phone system. The phone system shut down three different times that day. There were times when other Tucson telephone users could not even get a dial tone. TAPR sold around 650 TNCs over a period of two days.

Fig. 2-4. Packet radio newsletters and cover articles that have been published over the past several years.

TAPR also sold the rights to the TNC-2. However, this time the cost was $5,000 plus a royalty. TAPR felt the risk involved in a commercial company producing packet radio equipment had come down enough that the companies would be willing to put up more money. Today, the TNC-2 is sold by a variety of commercial companies. (See Fig. 2-5.)

At the end of 1985, there were approximately 10,000 TNCs in existence in North America.

The fifth ARRL Amateur Radio Computer Networking Conference was held on 9 March 1986 in Orlando, Florida. (See Figs. 2-6, 2-7, and 2-8.)

Fig. 2-5. The Pac-Comm TNC-200 circuit board.

At the 1987 Dayton Hamvention, Hank Oredson was presented with that year's Technical Excellence Award for his development of the WORLI packet bulletin board system. During 1987, packet radio began to really grow in other countries. At the end of 1987, it was estimated that well over 20,000 TNCs were in existence. By the summer of 1988, there were estimated to be around 35,000 packet operators.

Fig. 2-6. Lyle Johnson WA7GXD speaking at the Fifth ARRL Amateur Radio Computer Networking Conference. (Photo courtesy of FADCA.)

32 The History of Amateur Packet Radio

Fig. 2-7. Gwyn Reedy W1BEL at the Fifth ARRL Amateur Radio Computer Networking Conference. (Photo courtesy of FADCA.)

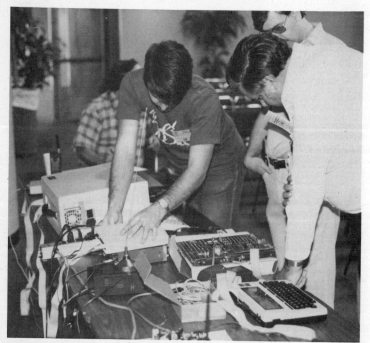

Fig. 2-8. Some of the packet equipment on display at the Fifth ARRL Amateur Radio Computer Networking Conference. (Photo courtesy of FADCA.)

Fig. 2-9. The two VADCG TNCs (original and TNC+) with manuals.

TAPR is involved in several projects at the present time. High on the list is a network node controller to allow for advanced networking. Work is also being done in the area of high-speed modems. The VADCG is still around and kicking. An improved VADCG TNC named the TNC Plus (Fig. 2-9) is available which supports three protocols: AX.25, V-1, and V-2. V-2 is the VADCG's new protocol. The original VADCG protocol has been named V-1.

CONCLUSION

There were, undoubtedly, many other important events, individuals, and groups that have contributed greatly to the present-day state of amateur packet radio that are not mentioned in this chapter; I apologize to those I missed. This is an attempt to summarize the major events that have occurred during the evolution of present day amateur packet radio.

There are still many important events to come in the continual evolution of amateur packet radio, and there is no reason you cannot be a part of them. There is much more to be done. The chances are that if you are even moderately active on packet, you cannot help but contribute in some form.

The next chapter discusses the design and function of the TNC along with other hardware components in a packet radio system.

Chapter 3

Amateur Packet Radio Hardware Systems

This chapter covers the hardware components of a packet radio station. Starting with a thorough description of the TNC, the chapter continues with discussions of modems and digital modulation. The chapter closes with a look at the transceiver requirements for amateur packet radio operation.

TNC

The *Terminal Node Controller (TNC)* is the heart of a packet station. The TNC serves as an interface between a user's terminal and the packet radio network. The TNC is also known as a *packet controller,* a *Packet Assembler/Disassembler (PAD)*, or a *Frame Assembler/Disassembler (FAD)*. The TNC is responsible for organizing and controlling the transmission and reception of data across a packet radio network. As you may recall from Chapter 1, there are two kinds of TNCs: software-based and hardware-based. Some TNCs are a mixture of the two, with some functions being handled by hardware and others by software. Today, however, the name TNC usually applies to only a hardware system. A collection of various TNCs is shown in Fig. 3-1.

Software-based TNCs

The best known software-based TNC is the Richcraft system. It is a complete packet system (with the exception of modem and transceiver) written for use on microcomputers based on the Z-80

Fig. 3-1. Assorted TNCs. Clockwise from the bottom: the original VADCG TNC, the Pac-Comm TNC-200, the Kantronics KPC-2, the AEA PKT-1, the Heathkit HD-4040, and the VADCG TNC+. The disk contains the Richcraft AX.25 TNC software system.

microprocessor, specifically the TRS-80 Models I, III, and IV. (For more information on the Richcraft system, see Chapter 7.)

In theory, a software-based TNC has several advantages over a hardware-based one. No additional hardware beyond the microcomputer and modem is needed, so the total system cost is reduced. Total power consumption is also reduced because the TNC software does not consume any extra power beyond that normally required by the computer. A software-based system makes maximum use of the available resources.

But a software-based system does have several crippling disadvantages. It is machine-dependent, so the software is limited as to what computer configurations it will work with. It is also expensive and time-consuming to develop software for a range of machines. With the wide proliferation of microcomputers, it would be all but impossible to develop software for each one. A software-based packet system puts stringent demands on the computer system. Since tasks must be carried out within very tight time frames, very efficient design criteria for both the software and computer must be met. Some computers simply lack the capability to handle a software-based TNC.

By using a software-based TNC, you restrict yourself to the user interface provided within the software. The user interface is your access to the packet system. Commands and other information are entered and received using the user interface, and if it is not convenient to use, you will either get used to it, suffer with it, or attempt to modify the software. More user interface options are available for hardware-based TNCs.

For these reasons, software-based TNCs have not become very popular. However, they do work and provide an inexpensive means to get on packet provided you have a compatible computer and modem available.

Hardware-based TNCs

Hardware-based TNCs are the mainstay of the packet marketplace. Many manufacturers have jumped on the packet bandwagon, and now there are numerous hardware-based TNCs available that vary in price and performance. However, one TNC has established itself as the *de facto* standard: the TAPR TNC-1. The TAPR standard was reenforced and expanded by the TNC-2. As explained in the previous chapter, TAPR had a fundamental role in the development of packet radio in the United States and worldwide. Until the TNC-1 was discontinued by TAPR, over 2500 units were sold and for the past few years they have set the standard of TNC performance.

The TNC-1 contains many features not found on other present day TNCs, such as a parallel status port and a wire wrap area on the circuit board for customized prototypes. It is an experimenter's board, but it also set the stage for "plug and chug" application-type operating. Its user interface has become a standard, and because of TAPR's liberal licensing agreements, many commercial TNCs are direct copies of TAPR's design.

A hardware-based TNC is actually a microcomputer system. As you will soon see, it contains the same basic components as a microcomputer. While different ICs may be used in different designs, the basic functions remain the same. No matter how many extras are added on, a TNC must have several standard capabilities in order to function properly.

Asynchronous I/O. All hardware TNCs must contain some sort of (I/O) capabilities for terminal communications. This is usually in the form of a serial communications port conforming to the RS-232 standard. (More information on the RS-232 standard can be found in Appendix B.) Through the I/O port(s), the TNC accepts data and commands and sends received data and status messages.

The RS-232 serial I/O port has become the standard for most peripheral communications. A peripheral is any component that is added on to a computer system such as printers, modems, and of course, TNCs. In serial I/O, the information in the form of bits is transferred serially one bit at a time over a data line. (See Fig. 3-2.) Other lines are usually included for control and to carry status information. Almost all microcomputers and terminals provide for RS-232 communications. When connecting a peripheral (such as a hardware TNC) to a terminal or computer using a RS-232 interface, a 25-wire ribbon cable is usually used for runs under 50 feet. The ribbon cable is usually terminated on each end with a DB-25 connector which is the standard connector in RS-232.

Parallel I/O is also used for connecting peripherals to computer equipment. Today it is used almost exclusively for connecting printers. In parallel I/O, the information bits that make up each piece of data are sent at the same time over individual wires. (See Fig. 3-3.) For example, if each piece of data is represented by one byte (8 bits), then 8 separate wires would be used. There are other wires added for carrying control and status information between the computer and the peripheral. Parallel I/O is not used much in most TNC designs.

Parallel transfer is used internally in most computer systems (and TNCs) to transfer information between its components. For example, information is transferred between the processing unit and memory in parallel. Thus, a means of converting from parallel to serial and back again is needed to communicate with the outside world

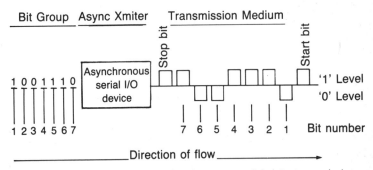

Fig. 3-2. A diagram showing asynchronous serial data transmission.

38 Amateur Packet Radio Hardware Systems

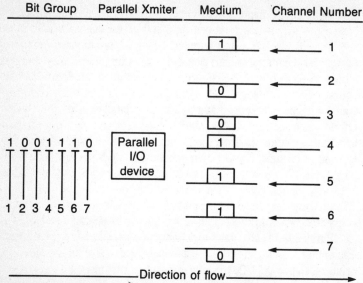

Fig. 3-3. A diagram showing parallel data transmission.

through a serial port. This is accomplished by the use of a Universal Asynchronous Receiver and Transmitter (UART).

The UART takes information fed to it in parallel format and sends the information out in serial format. In reverse, the UART accepts serial data and converts it to parallel format. UARTs are found in almost all serial communication I/O circuits.

Memory. Memory is another component that all TNCs share. *Random Access Memory (RAM)* is used for storage of short term information such as variable parameters and as a buffer for received and sending frames. Since RAM loses its contents when power is removed from it (volatile memory), some form of backup is usually provided in the form of a small battery cell on the circuit board to maintain power to the RAM at all times. One other method of backing up the RAM is represented by the TNC-1, which uses a special kind of non-volatile RAM for storage of valuable information. Most TNCs come equipped with a minimum of 16K (16,384 bytes or 131,072 bits of information) of RAM.

Read Only Memory (ROM) is another form of memory found on most all TNCs. The ROM is actually *Programmable Read Only Memory (PROM)* which is programmed, or "burned in," with the permanent programs needed to run the TNC. These programs consist of the user interface, the protocol(s), the calibration routines, and any other programs necessary to use the features of the TNC. Some TNC manufacturers put user information such as call-sign and VADCG number in ROM. Communications with the TNC is through the user interface. It accepts commands and displays status information.

The protocol(s) implementation programs which are contained in ROM usually consist of the AX.25 protocol and the VADCG protocol. These two protocols along with a discussion of protocols in general are covered in the next chapter. The protocol contains the rules to keep track of each frame and controls the transmission of frames.

The amount of ROM that comes with a hardware TNC varies depending on the number and complexity of programs included but is usually 8K to 32K. ROM is an ideal medium for storing long term information, and it can be easily removed for upgrading should a new version come out.

HDLC. So far you know that the information or data to be transmitted comes from the I/O port and that the program that controls the transmission of the frames is contained in ROM. But where does the information that comes into the TNC via the I/O port for transmission get put into frames and actually sent? This is the job of the *HDLC*. HDLC stands for High-level Data Link Control, and one of its functions is to format data into frames for transmission. It generates the *Frame Check Sequence (FCS)* of outgoing frames. If you recall from Chapter 1, the FCS is the root of packet's error-checking. The HDLC also disassembles and checks the FCS of received frames. Some TNCs handle the HDLC functions in software, while others use a separate IC designed especially for HDLC functions.

CPU. Just as the TNC is the heart of the packet station, the CPU is the heart of the TNC. It manages the operation of all other components and serves as a clearinghouse for all data transferred between components. The CPU follows the instructions programmed in ROM.

The speed with which the CPU performs its tasks is controlled by the system clock. The system clock generates all timing signals for the CPU and other components. Along with the capabilities of the components, the system clock puts an upper limit on how fast the TNC can work.

Radio I/O. After the HDLC has assembled a frame for transmission, it is sent to the modem. The modem circuit is usually, but not always, included on the TNC circuit board. If it is not included on the board, an I/O connector is provided on the TNC to interface with a modem. TNCs with on-board modems should have an external modem connector or some way to bypass the on-board modem should it be necessary to add an external modem. This feature is desirable to accommodate greater filtering if needed or a change in the standard modem specifications used for packet radio.

TNCs with on-board modems usually include a calibration program so it is easy to calibrate the modem without extra equipment. An on-board modem is a great convenience; however, a TNC without a way to easily bypass the on-board modem can cause problems if the modem standards change significantly. Modem standards and digital modulation techniques are covered later in this chapter.

While information usually comes to the TNC in asynchronous format from the terminal, the information in packet form is sent to the modem in synchronous format. In synchronous communications, there are no start and stop elements surrounding each bit group. Rather, the bit groups are combined into one long *bit stream* and sent as a whole. (See Fig. 3-4.) Because the sending speed and length of bit groups are predefined, the receiving station is able to identify the individual bit groups without the use of start and stop elements. However, the receiving station must know when the bit stream starts and when it ends. Thus, a special character (called a *flag*) is usually put at the beginning and end of each bit stream.

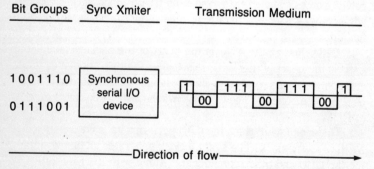

Fig. 3-4. A diagram showing synchronous serial data transmission.

Power Requirements. The power requirements of hardware-based TNCs vary from a present day low of 25 mA to 1 amp and up at 9 to 12 Vdc. If you are planning to operate portable packet, current drain is an important criteria to keep in mind. Sometimes the power consumption of a TNC can be reduced by replacing regular *Metal Oxide Semiconductor (MOS)* chips with their *Complementary MOS (CMOS)* equivalents where possible. CMOS chips draw much less power. Unlike the original TAPR TNC-1 which included an ac power system, most TNCs today run from 12 Vdc.

TNC SYSTEM TRACE

Now let's do a trace of hardware TNC based on the knowledge we have of its components. Assume that a terminal of some sort is connected to the I/O port, that a transceiver is connected to the modem, and that the modem is either on-board or interfaced in some way to the board. (See Fig. 3-5.) Some introductory information about the formation of a frame is given. (More detailed information can be found in the next chapter.) The first trace is the path of an outgoing packet through the system.

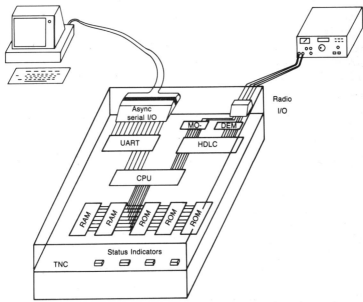

Fig. 3-5. A "blowout" of a hardware TNC illustrating the various components.

The information to be transmitted to the receiving station is typed on the keyboard of the terminal. The information goes through the I/O port into the TNC and is directed by the CPU to a RAM buffer as controlled by the software in ROM. Once the maximum length of the information is reached (usually 128 to 256 characters), or a send frame command is received from the terminal, the information is sent to the HDLC along with other control data. The HDLC combines the control data and the information into a frame and then computes the *Frame Check Sequence (FCS)*. The FCS is added to the end of the frame.

The completed frame is then sent to the modem where the individual bits from the HDLC modulate the signal sent to the transmitter. The transmitter then transmits the modulated frame to the receiving station (usually in less than a second provided the link between the two stations is clear and reliable).

The receiver at the receiving station receives the modulated frame and sends it to the modem. The modem demodulates the frame and sends the bits to the HDLC. The HDLC then disassembles the frame into its component parts (control data and information), provided the FCS check is good. The information is then routed to the terminal I/O port or RAM if the port is busy, as directed by the CPU following the program in ROM. The information is sent through the I/O port to the terminal where the information is displayed exactly as it was sent.

Each frame is sent this way. Whether you are transmitting a short note to your friend down the street or a large file across town, it all goes one frame at a time.

MODEM

The modem is an integral part of any radio-based digital communications system. In packet radio, the modem interfaces between the TNC and the transceiver. There are many different types and configurations of modems.

A modem inputs bits from a digital device (in this case a TNC) and modulates the transmitter with an audio signal (a sine wave) so that the information can be transferred via radio frequency. The modem can also receive modulated signals from the receiver and output bits whose state vary according to the content of the received signal.

There are several different modulation (and reciprocally demodulation) methods in use today. Only the frequency, amplitude, and phase components of a sine wave (see Fig. 3-6) can be changed for the transmission of information between modems. The forms of modulation covered in this chapter are *Frequency Shift Keying (FSK), Audio Frequency Shift Keying (AFSK),* and *Phase Shift Keying (PSK)*. The type of modulation used determines such factors as bandwidth at a given baud rate and the type of transceiver necessary. We'll come back to the subject of modulation after we take a look at baud rate in greater detail.

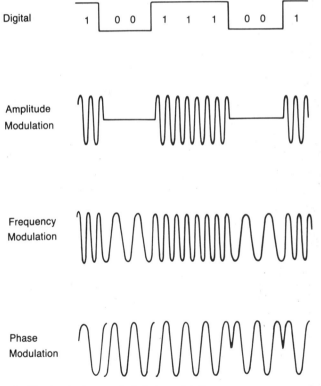

Fig. 3-6. The basic forms of digital modulation.

BAUD RATES

Earlier, the baud rate was generally defined as the number of bits sent per second. This definition is not entirely accurate in all cases. The baud rate is usually equal to the *Bits Per Second (BPS)*, and the two terms are often used incorrectly as synonyms. A baud is a measure of the signaling rate. A baud is actually the number of discrete events per second. For example, a 300 baud signal consists of 300 discrete events per second, and if each event represents a single bit, the BPS equal 300 also. However, in some forms of modulation, each discrete event can represent more than one bit; thus, it is possible to have a rate of 600 or 1200 BPS from a 300 baud signal.

A discrete event is a transition from one level to another in the modulation scheme. For example, changing from a 1000 Hz signal to a 2000 Hz signal is a transition (or discrete event). If a demodulator were designed to convert a 1000 Hz signal into a 1 bit and a 2000 Hz signal into a 0 bit, we would have a two level modulation scheme. And the number of bits each discrete level represents would be 1. If we were to increase the number of frequencies to 4 (1000 Hz, 2000 Hz, 3000 Hz, and 4000 Hz) then we could represent 2 bits per event (i.e., 1000 Hz = 01, 2000 Hz = 11, 3000 Hz = 10, and 4000 Hz = 00).

It is possible to represent the above in the form of a mathematical equation: BPS = BAUD∗I. The number of bits sent per second is equal to the baud rate multiplied by the number of bits that each discrete event can represent. Another example: a 1200 baud signal with each discrete event capable of representing 2 bits has a data transmission rate of 2400 bits per second. More information is given later in the chapter on modulation systems that support multiple combinations per event.

Channel Characteristics

While we're on the subject of equations, this is as good a time as any to introduce two more that are useful when finding the data rate of a communications channel. The first equation is known as Nyquist's equation. It says that the maximum baud rate is equal to twice the bandwidth in Hz (Baud = 2∗Bandwidth). For example, a 1200 baud signal requires 600 Hz bandwidth.

The Baud = 2∗Bandwidth equation is actually a simplification of Nyquist's equation. It assumes that each baud represents a single bit. In the case of a baud representing more than one bit, Nyquist's full equation is BPS = 2∗Bandwidth∗Log2 (# of discrete levels). As you can see, as long as there is only one event per baud (2∗1), the

Log2(2) equals 1, and the equation becomes BPS = 2∗Bandwidth∗1. Since the BPS is equal to baud rate in this case, the equation can be simplified to Baud = 2∗Bandwidth as initially introduced.

But, Nyquist's equation assumes a communication channel free of noise, interference, and distortion. Because of this limitation, a communication channel will never obtain the exact characteristics obtained from the equation. It is useful as an approximation and gives an actual value for analysis even if it is very optimistic.

The next equation, Shannon's law, is somewhat more useful as it takes noise and interference into account. The amount of random and thermal noise on a channel is measured by the signal power to noise power ratio, more commonly called the signal-to-noise ratio. The signal-to-noise ratio is usually represented as S/N. The signal-to-noise ratio is included in Shannon's equation as follows: BPS = Bandwidth∗Log2(1 + S/N).

More on Modulation

As mentioned earlier, a sine wave can be characterized by frequency, amplitude, and phase. Phase is useful as a signaling device only if the frequency remains constant. Since a sine wave completes a full cycle in 360 degrees, the amount of lag that one signal has compared to another one can be used to indicate a transition or discrete event.

The difference between FSK and AFSK frequency modulation has to do with the method used to transmit the generated signal. FSK is classified as direct modulation, and AFSK is classified as indirect modulation. In direct modulation, the transmitter carrier frequency is shifted up and down to transmit the analog signal. In indirect modulation, the transmitter's carrier remains stable while an external signal is superimposed upon the carrier. It is the external signal which varies, not the transmitter's carrier.

AFSK is usually easier to implement with modern transceivers as all signal generation can be done externally of the transmitter. The audio signal can be simply fed into the transceiver's mic input. Since most transceivers do not include direct FSK ports, some modification is usually necessary to implement FSK directly. Regardless of whether the transmitter is AFSK- or FSK-controlled, the transmitted signal is the same.

In (A)FSK, the frequency of the sine wave is varied to indicate transmission of a binary 1 or 0. As mentioned before, the bandwidth required varies according to the signaling rate, so the minimum difference or shift between the two frequencies is dependent upon the signaling rate. If a signaling rate of 200 baud is used, the frequency pairs used to represent the binary 0 and 1 might be 1000 Hz and 1200 Hz. While it is possible to have a multi-level modulation scheme using FSK, it is not normally utilized. At present, AFSK is the means of digital modulation most used in amateur radio communications, packet included. However, that may soon change as we look for methods that allow us to transmit more information through less bandwidth and under poorer channel conditions.

Phase Shift Keying offers the benefits listed above and may see increased usage in the future in packet radio. PSK transmits information by changing the phase of the signal. If the phase is shifted between 0 degrees and 180 degrees, it is possible to transmit information with each phase degree representing a binary character. If we were to add more phase shifts, 90 degrees and 270 degrees, we could represent two binary characters in each phase shift. For example, 0 degrees = 10, 90 degrees = 01, 180 degrees = 11, and 270 degrees = 00; each pair of binary characters is called a *dibit*. Using the dibit, the BPS has been doubled while maintaining the baud rate.

It is easy to see why PSK could be one of the modulation schemes to be reckoned with in the future. If more phase shifts were added and some amplitude modulation were thrown in, the data rate could conceivably be quadrupled. *Quaternary Amplitude Modulation (QAM)* is such a form in which PSK is combined with amplitude modulation to encode 4 bits per transition. The one drawback to PSK is that it requires more complicated modem units than regular FSK.

AMATEUR PACKET RADIO MODULATION SCHEMES

Now that you have a better idea of how modems operate, we can discuss the modem standards in use today on packet radio. There are presently two different standards, one for VHF operation and one for HF operation. Soon, a third and even fourth standard may emerge for use on VHF and UHF. But before we get too far ahead of ourselves, let's digress a little.

In the early days of amateur packet radio, the late 1970s and early 1980s, packet experimenters were looking for modems to use with their systems. They had to be cheap, reliable, and plentiful. The modems also had to be simple; they were having enough problems with the TNCs alone. The modems had to have a fairly high baud rate and be easy to interface with the rest of the system. They chose the Bell 202 standard because it met all the above requirements.

Bell 202 modems are still found as surplus items and at hamfests for very low prices. The Bell 202 standard uses a 1000 Hz shift with mark and space tones at 1200 Hz and 2200 Hz. They can handle the standard 1200 baud rate of VHF packet radio, and are easy to calibrate and use.

The Bell 202 modem was implemented in packet radio for use with 2-meter FM transceivers. The modulation method used is AFSK at 1200 baud. Bell 202 modems are the type found on most on-board TNC modems. The Bell 202 modem usually uses a *phase lock loop (PLL)* demodulator.

A PLL demodulator works by utilizing a phase detector along with a *voltage-controlled oscillator (VCO)* in a feedback circuit. A direct current feedback voltage is generated proportional to the difference in frequency between the received audio and the VCO. This voltage changes as needed to adjust the VCO to the same frequency as that which was received. Therefore, the voltage will vary as the input audio frequency alternates between mark and space conditions. This varying voltage is then filtered and amplified to produce the required mark and space signals.

There are several disadvantages to using PLL demodulators. One is that they tend to lock onto the strongest signal in their lock range, often ignoring weaker signals that you might be trying to copy. Another disadvantage is that a PLL has no variable tuning indicator, so it must be tuned by ear in conjunction with a single LED which lights when a signal is tuned in. However, these disadvantages are not too important on VHF and UHF packet radio. The fact that the PLL tends to lock in on the stronger signal is useful if two stations transmit at the same time. The stronger signal might be received correctly, thus avoiding a complete collision. And since most packet activity on VHF and UHF takes place on pre-assigned fixed frequencies, tuning is not a major issue.

For HF operation, *shift filter*-based modems are usually used. The filters are tuned for particular shift frequencies and come in two different flavors: passive and active. A passive filter is a simple LC resonant circuit whose band-pass is pre-adjusted to pass the necessary frequency pairs. An active filter system (either transistors or ICs) may utilize feedback filters or switched capacitor filters to provide the desired band-pass characteristics.

There is an increasing amount of packet activity on the HF bands. The 10, 20, 30, 40, and 80 meter bands are being used for transcontinental and DX communications. There are HF Gateway stations which allow for input on VHF using the Bell 202 standard and output on HF. Thus, a modest station or one without room for an HF antenna can take advantage of the increased range of HF. More information on operating packet on HF is given in Chapter 6.

Because of the characteristics of HF operation, different modem configurations have been developed for packet transmission. 300 baud (A)FSK with a 200 Hz shift is the usual standard. If AFSK is being used, the LSB (Lower Sideband) is chosen. On 10 meters, the Bell 202 standard is sometimes used.

Although Bell 202 on VHF is the mainstay of packet activity and HF is being increasingly utilized, other forms of modulation have been under development and will see increasing usage in the future due to their obvious advantages. 400 baud FSK is being used for amateur satellite activity. 9600 baud PSK and FSK is under development for high speed linking, and very high speed (56,000 BPS) radio modems are under development for real time digitizing of voice and video along with higher speed linking.

RADIOS

In order for a transceiver to work well on packet radio, it must meet several requirements. A fast turnaround time from transmit to receive and vice versa (T/R time) is important for efficient packet operation. A slow T/R time will reduce the throughput of the packet network. A T/R time of 50 milliseconds is fine and up to 100 ms is tolerable. Radios with T/R times from 5 to 15 ms are ideal. Since the transceiver will most probably be left on for long periods of time, good stability is a necessity. Most modern rigs do not have a problem in this area.

The transceiver's bandwidth must be capable of passing the audio tones used in modulation. Low distortion is another desirable feature. The relative amplitude of modem tones should not be altered significantly by the radio. Changes in phase are tolerable as long as the change is linear (all frequencies are shifted the same amount).

Don't let these requirements worry you too much. Most any recent transceiver is adequate for packet operation. On VHF FM, a handheld will do. HF operation is somewhat more critical of transceiver performance; a modern synthesized rig is recommended.

Antenna requirements are discussed in Chapter 5. Pre-amps, receive diversity, special filters, and other considerations can be implemented as needed. But, the chances are good that whatever set-up you presently have can be used for packet operation without any major changes.

CONCLUSION

This chapter covered the different hardware components of a packet radio station. Starting with the differences between a software-based TNC and a hardware-based TNC, it discussed the components of a hardware-based TNC and their functions, different types of I/O, and the various schemes of digital modulation along with the common ones utilized on the VHF and HF bands. The chapter ended with a look at transceiver requirements for effective packet operation.

The next chapter covers the second half of the amateur packet radio system: networking and protocols.

Chapter 4

Networking and Protocols

This chapter covers two related areas that are an integral part of packet radio: networking and protocols. These are not items which can be purchased through a mail order company or breadboarded like hardware. Networks and protocols are concepts which, when implemented along with hardware systems, result in a viable packet radio system. All three components (hardware, networks, and protocols) work together, and a failure in one can result in the failure of the entire system. As in the case of hardware, there are many different kinds of networks and protocols, each favoring certain conditions. This chapter introduces networks and protocols, explores the various options of each, and describes the common systems in use today.

NETWORK BASICS

A single packet station is useless for communications. In order to communicate, two or more stations are needed. While this problem is not encountered by much today, in the early days of amateur packet radio many users would rush to put their station together only to find that there were no other packet stations in their area. When there are two or more packet stations within communications range of each other, a *network* is formed. In terms of digital communications, a network can be defined as a collection of communication devices linked together so that one station can communicate with any other station in the network. The difficulty with networking is deciding on and implementing a system which allows for maximum flexibility and throughput while minimizing complexity and cost.

In the simplest case, a packet network consists of a few stations within direct communications range from each other on a single frequency. (See Fig. 4-1.) A more complex network involves digipeating (simplex packet repeaters) to extend a station's communication range and gateways for accessing stations with different capabilities (such as those on another frequency or using another modem configuration) as shown in Fig. 4-2.

However, this system is not ideal in many respects due to congestion, range limitations, and other problems. Before more advanced work can be done dealing with packet networking, additional work has to be done in the area of protocols. The present day packet system has stretched the current protocols to their limit, and much work is being done in the area of developing protocols (more on this later in this chapter). The last chapter in this book focuses on the future of amateur packet radio and the subject of advanced networks is discussed.

MULTIPLEXING

Since packet operation today occurs on agreed-upon single frequencies, a method of allowing stations to access the frequencies in an orderly manner is necessary. If each station transmitted whenever it wanted to, collisions and other problems would occur. There

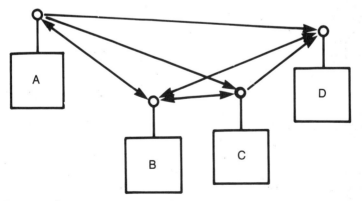

Fig. 4-1. A diagram of a simple network in which all stations can communicate with each other directly.

52 Networking and Protocols

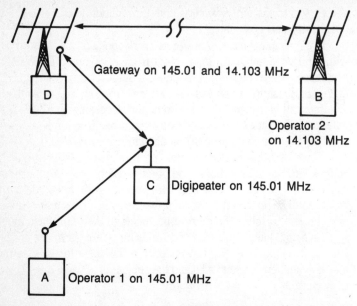

Fig. 4-2. A diagram of a more advanced network in which a digipeater (station C) and HF gateway (station D) are utilized to allow station A on 2-meters to communicate with station B on 20-meters.

must be a way to allow each station to use the frequency (channel) without interfering with other users.

The method used is known as *multiplexing*. Multiplexing allows a group of users to share a communication medium. In its ideal form, each user is not aware that he or she is actually sharing the channel. It should seem, from the user's perspective, the he or she is the sole user of the channel. Two forms of multiplexing immediately concern amateur packet radio operators: *Time Division Multiplexing (TDM)* and *Frequency Division Multiplexing (FDM)*.

FDM

FDM allows each transmitting user to have a separate channel for communications. A good example of this is the radio stations on your FM stereo. Each station has its own frequency and occupies it continuously. In an active two-way communications system such as amateur radio, it would be wasteful to assume that each station could have its own frequency as packet stations need only use the channel for brief periods of time. There are usually a set number

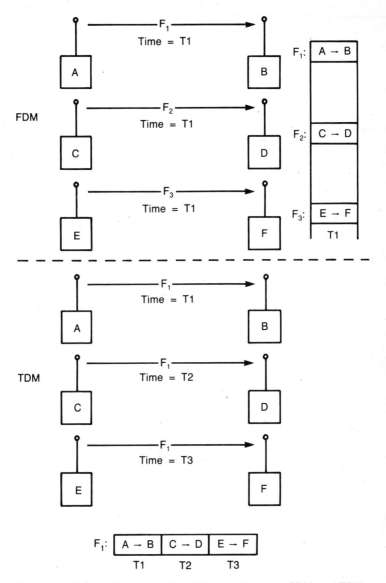

Fig. 4-3. A diagram illustrating the difference between FDM and TDM. In FDM, all three transmissions occurred simultaneously on separate channels. In TDM, the three transmissions occurred sequentially on a single channel.

of frequencies allocated for communications, and the user selects one before initiating communications.

Usually, once a station has begun to communicate over a certain channel, it remains on that same channel for the duration of the communications session. This is known as *static FDM* because the stations do not switch between different frequencies during a connection. Pure FDM operation does not provide a very versatile network.

TDM

With many users sharing a common channel, we turn to TDM to allow each user to access the same channel without interfering with other users. In TDM, each station transmits one after the other while users with no traffic stand by. Thus, the frequency is allotted by time to users with traffic to send; one station transmits for a time period, and when it is through, another station transmits. Let's discuss how a station knows when it is OK to transmit.

There are three popular systems in use which control the access of individual stations to a communications channel. These are *random access, polling,* and *token passing*. Random access is used in amateur packet radio.

In a polling system, a master station asks each station on the network channel individually if they have traffic to send. If so, they respond positively and then transmit as the channel is kept clear since no other station has been cleared by the master station to transmit. Another form of TDM similar to polling systems is *token passing*. In a token passing system, a single electronic token (a special binary sequence) is passed from station to station until it arrives at a station which has traffic to send. The station holds onto the token and transmits. The frequency is kept clear because only the station with the token is allowed to transmit. When the station has finished transmitting, it passes the token to the next station on the network. Depending on the configuration of the network, individual stations may communicate with each other directly or via the master station.

One reason polling systems have not become popular in amateur packet radio is that they require a master station, which must have a fairly powerful computer and reliable communication to all network users to keep track of the users and their status. Because of the nature of amateur packet radio, users tend to drop in and out quickly and the radio links between stations vary in quality. In order for a polling network to work effectively, network conditions must be regimented beyond what most amateurs can provide due to the increased cost and high reliability needed. Another reason polling has not been proposed more for amateur packet radio is the amount of overhead required. The concept of overhead will be discussed later, however; a quick definition is the amount of information that must be added to the basic data in order to route the data to its destination.

The system that is in use in amateur packet radio today is random access. In a random access system, each individual station accesses the network as it sees fit under a defined set of rules. Each station must be able to determine if the channel is fit to be used; i.e., clear. In amateur packet radio, the method used is known as *Carrier Sense Multiple Access with Collision Detection (CSMA/CD)*. Each station monitors (senses) the channel and when the station has traffic to send, it checks to see if the channel is clear. If the channel is clear, the station transmits. A successful transmission is indicated by the reception of an acknowledgement from the destination station. If the channel is not clear, the station waits until the channel is clear and then transmits. If two or more stations transmit at the same time, a collision may occur. If a collision does occur, no acknowledgement is received by the stations involved in the collision, so each waits a random length of time and then attempts to retransmit. One station has a shorter random wait than the others and captures the channel first, thus avoiding another collision (at least with the stations originally involved in the collision).

Packet radio uses both FDM and TDM in order to permit many users to transmit and receive simultaneously. The channel (frequency) selected by FDM affects both the range and speed of data transmission. For example, channels in the 20-meter band have a large range but limited speed, and channels in the 2-meter band have a limited range but support much higher speeds.

TDM allows multiple users to share the same channel. CSMA/CD is used to implement TDM. One important point to keep in mind about CSMA is that all the stations on the channel must be within hearing range of each other for it to work for all stations.

SDM

This introduces a new kind of multiplexing: *Space Division Multiplexing (SDM)*. SDM comes about because all stations cannot hear all other stations. SDM becomes increasingly applicable as higher frequencies with limited ranges are used. It lets two operators use the same radio channel at the same time without interfering with each other. For example, a station in California can transmit on 145.01 MHz at the same time as a station in Pennsylvania, and because the two signal paths do not cross, they can coexist without any problems. In essence, they are not on the same network. SDM can result from a variety of causes including propagation, radiation patterns, and physical obstructions. In the VHF and UHF bands, such effects are fairly constant and can be predicted quite easily.

AMATEUR PACKET RADIO MULTIPLEXING MODEL

Combining the effects of FDM, TDM, and SDM, it is possible to develop a packet radio multiplexing model. Imagine a layer cake with thin layers. Each layer represents a different frequency, and the area of each layer is proportional in size and shape to the total geographical area of the network. The cake has a large number of layers, each layer representing a packet radio channel. Individual stations operating at certain frequencies are represented on the appropriate layer by points placed at the proper geographical location. The range of a station is plotted around that station's location. The size and shape of the range (coverage) plot depends on the frequency used, the amount of power the station is transmitting, the gain and directionality of the antenna, and propagation conditions. Differences in transmit and receive coverage are not taken into account for the sake of simplicity.

If the range plots of two stations overlap on the same layer, they can communicate directly. If the plots do not overlap, a digipeater or gateway may have to be used to bridge the gap. If the range plots of two or more groups of stations (networks) do not overlap, the groups of stations may operate as independent networks concurrently without interfering with each other. If the range plots between two groups of stations do overlap, they are not independent and TDM must be implemented to share the channel. This model can be applied to local as well as wide area networks by simply extending the total area that each layer of the cake represents. See Fig. 4-4 for a graphic representation of the above concepts.

Amateur Packet Radio Multiplexing Model 57

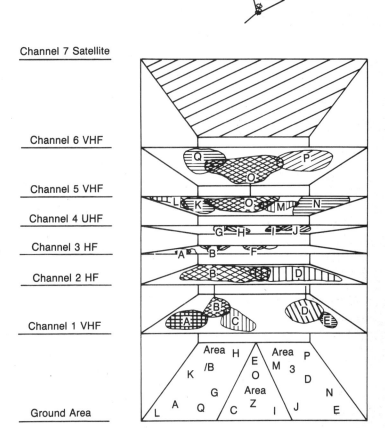

Fig. 4-4. A model representing the present-day amateur packet radio multiplexed network. Stations B, D, and O are gateways which allow stations on one layer to communicate with stations on another layer. By spreading the total number of stations over several channels, the congestion on a single channel is reduced. Notice the wide coverage of the satellite channel.

PROTOCOL BASICS

Now that we've discussed the basic networking concepts and a physical picture of the packet radio network has been presented, we can take a look at the intelligence behind the systems that are implemented to make packet radio work. How do individual stations know how to communicate with each other? What happens if data is sent but for some reason is not received? What if the data arrives garbled? How does a station know if received data is meant for it? How do digipeaters know to retransmit certain data but not others? The answers to these questions and many more are found in the protocol used.

Protocols define how data is to be packaged and what actions are to be taken under certain conditions along with when the actions are to be executed. The ultimate goal is to get data from the originating station to the destination station as quickly and efficiently as possible with no errors. The general steps involved in communicating using packet protocols are discussed in the following section.

Assuming that a station can access the network and reach other stations, it must let a station know that it wants to communicate. In packet radio, this is known as a *connect request*. If the selected station is available, it acknowledges the connect request and the two stations are connected. Once the stations are connected, the information they send to each other is received error free. And when the two stations are finished communicating, they must disconnect from each other and be ready to connect to other stations.

All of these processes and many more are handled by *protocols*. A protocol is a predefined series of steps which are followed to accomplish something. For example, at a formal dinner, protocol may define what is to be worn, where people are to sit, what fork is to be used, etc. A simple example which illustrates how a random access packet radio protocol works is a normal 2-meter FM phone contact.

First, you call the station you wish to communicate with. (AA3F, AA3F this is KR3T. Do you read me?) At all times throughout the communication, you first listen to see if the frequency is clear before transmitting. You keep calling until he responds or you get tired and give up. Assuming he responds, you have now established a connection. (KR3T this is AA3F. Go ahead.) You would then transmit the information you wish to communicate. (AA3F this is KR3T. Meet me at the mall in 5 minutes.) Notice that you send the receiving and sending station's callsigns so that AA3F knows this message is for him. (The FCC likes it too!) If AA3F responds, or acknowledges (KR3T this a AA3F. Roger.), then you know the message was received. If AA3F does not respond within a reasonable length of time or asks KR3T to repeat the message, you retransmit the message again and again until he acknowledges it or you get tired and give up. The connection is ended by sending a disconnect request. (AA3F this is KR3T. 73.) AA3F responds. (KR3T this is AA3F. 73.) You have just ended the connection and are now free to communicate with another station if you wish.

A good example of a polling protocol is a 2-meter repeater net. In a repeater net, the repeater is used as the primary station. All individuals transmit to the repeater and the repeater relays the signal to the other users. One user is designated *Net Control Station (NCS)*. The NCS acts as the equivalent of an amateur radio central node or master station.

The NCS is responsible for initiating the net by calling for stations to join in (the equivalent of logging onto a central node). The NCS maintains a list of those stations currently in the net. After the net has begun, new stations must call the NCS and request to be included.

Referring to the list of stations currently logged into the net, the NCS calls each station individually and asks if they have any traffic to send (the equivalent of polling). If the station does have traffic, the channel remains clear while the polled station transmits. Once the polled station has finished transmitting, or if the polled station does not have any traffic, the NCS polls the next station on the list.

The NCS assumes all stations on the list are operating unless the station does not respond to a poll or requests to be excused from the net. In both cases, the station is deleted from the list. If one station is considered to be of more importance than the others (i.e., emergency traffic), it may be polled earlier and more than once in each cycle through the list.

The same basic systems apply to packet radio. A connection is established, information is transferred (sending information again when it is not received properly), and the connection is cancelled. Keep in mind that these are very simple generalizations. The protocol must be able to determine when information is received incorrectly, keep track of the connection's status, translate data, assure device compatibility, and much more. A more detailed look at how protocols are organized and the protocols in use in amateur packet radio follows.

OSI/RM

Any network, packet or not, is made up of a multitude of different components and functions (i.e., terminals, codings, voltages, error-checking, connecting, relaying, and disconnecting). Networks can become very complex as their capabilities increase, and numerous approaches are possible. To alleviate some of the complexity, the *International Standards Organization (ISO)* developed a reference model for networks. The model is known as the *Open Systems Interconnection Reference Model (OSI/RM)*. The OSI/RM is designed to facilitate the exchange of information between systems. A system can be as simple as a current loop teletype and as complex as a worldwide network. Open systems are systems that are available for communications, like a packet radio station. The OSI/RM separates network functions into different levels based on their purpose.

Each OSI/RM level transfers data between the level directly above and directly below it in the level hierarchy. The point where data is transferred between levels is called the *interface*. Data originates at the highest implemented level and is passed down serially through each level where the data is processed by that level's protocol until it reaches the lowest implemented level. When the data is received, the path of the data is reversed and the data is sent up the levels where each level removes whatever additional information was added by that level's equivalent at the sending station. By the time the data reaches the level from which it was originated, it looks exactly the same as when it was entered into the network.

Each level operates independently of the other levels. The only exchange of information between levels occurs at the predefined interface points. Each level has a protocol associated with it. One level's protocol can be changed without affecting the rest of the levels. The set of levels and associated protocols form the network architecture.

The OSI/RM is the basis on which the amateur packet radio network is being developed. Its flexibility and structure are a great help in maintaining compatibility between packet systems. The OSI/RM is divided into seven levels (layers). Each level is responsible for particular tasks and has been assigned a name representative of its function.

The seven OSI/RM layers are: physical, data link, network, transport, session, presentation, and application. (See Fig. 4-5).

Level 1 is the physical layer. It is responsible for the transparent transmission of bit streams across the physical interconnection between systems. The physical connection can be operated in either simplex, half duplex, or full duplex. The bits must arrive in the same order in which they were sent. Specifications for the physical layer include mechanical (such as plugs and dimensions),

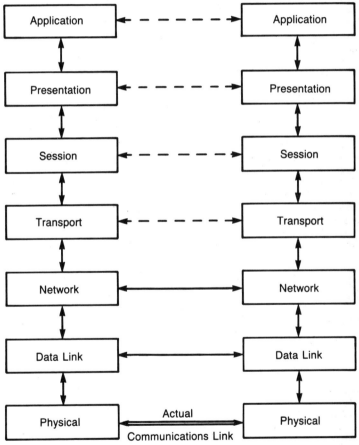

Fig. 4-5. The levels of the OSI/RM. The solid arrows indicate layers that must be implemented at each node. The layers indicated by dashed arrows must be implemented only at the originating and destination nodes.

electrical (such as voltage levels), and procedural (such as rules and sequences).

Level 2 is the data link layer. The data link layer is responsible for shielding the higher levels from the characteristics of the physical layer. It provides reliable, error-free transmission of data. The data link layer should contain some form of error detection and correction. The data link layer must also be independent of the data sent; it may not alter the data in any way. It must accept data and break it into segments for transmission. When the segmented data is combined with protocol information, a frame is formed. The frame must be *delimited* (i.e., allow for recognition of the beginning and end of the frame), and the frame must also be transparent (looked at only as a series of bits).

The frame must be checked for errors upon reception and sent again by the last station that has a copy of the frame if an error was found. The frames must be delivered in the same order that they were sent. The data link layer is responsible for a great deal. The standard level 2 protocol is *High-level Data Link Control (HDLC)*. A subset of the HDLC is used in most amateur packet radio data link layer protocols. The data link layer utilized in packet radio is discussed in detail later in the chapter.

Level 3 is the network layer. It is responsible for providing transparent transfer of all data submitted by the transport layer (level 4). The network layer completely relieves the transport layer from any concerns about the way in which the communicating systems are connected. The systems may be connected point-to-point (direct) or may have many systems (nodes) in the path. The network layer must provide the necessary routing functions necessary to get data from one system to another in the network; each system may act as a relay. The actual routing methods are not covered in the OSI/RM. Level 3 of the X.25 standard is one standard protocol for the network layer.

Level 4 is the transport layer. It is responsible for arranging the information in the correct order in the event the packets arrive out of sequence. The transport layer is only concerned with communications between the originating and destination stations and not with any relay stations that might be utilized by the network layer.

Level 5 is the session layer. The session layer is responsible for initiating and terminating communications between stations on the network.

Level 6 is the presentation layer. The presentation layer is responsible for data transformation (e.g., converting ASCII to Baudot), data and display formatting (for example, a graphics terminal communicating with a hardcopy teleprinter), and syntax selection. If the two communications systems are using incompatible devices, the presentation layer handles the conversions necessary for the two devices to transfer data.

Level 7 is the application layer. The application layer is responsible for the proper operation of application-entities. Application entities can be defined as user-oriented software. Any programs or computer functions controlled by the connected system would be located at the application layer and fall under the control of its protocol(s).

AMATEUR PACKET RADIO PROTOCOLS

This section describes the levels of the OSI/RM as they relate directly to amateur packet radio. The levels currently implemented in amateur packet radio in the United States are the physical layer, the data link layer, and rough forms of the network layer. Other levels and protocols currently under development are also discussed.

Physical Layer

The physical layer is fairly well standardized. The Bell 202 and Bell 103 are the most widely used modulation standards. The RS-232C asynchronous serial interface is another physical layer standard. Other modulation standards will emerge as high speed modems and new modulation schemes are developed.

One area of the physical layer regarding the transmission of data which is specified in the physical layer is *encoding techniques*. The encoding technique defines the format of the modulated signal. In ordinary RTTY, as well as present day amateur packet radio, a bipolar format is used. In bipolar keying, two different signaling levels are utilized for the representation of binary 1 and 0. Bipolar keying is an improvement over unipolar keying in which a single tone is used to represent a binary 0. (See Fig. 4-6.)

There are several forms of bipolar keying. The one that is used by regular Baudot RTTY and AMTOR is known as *Non-Return to Zero (NRZ)* or *NRZ-Level (NRZ-L)*. In NRZ, a binary 1 is represented by one level or tone, and a binary 0 is represented by another level. Thus, if the data being transmitted is the binary

64   Networking and Protocols

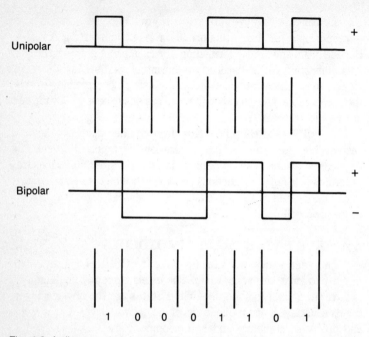

Fig. 4-6. A diagram showing the differences between unipolar and bipolar keying techniques.

grouping of 1001110, the signal switches (transition) from a level representing the binary 1 to the level representing the binary 0, remains at the binary 0 level, switches back to the binary 1 level, remains at the binary 1 level for the next two bits, and then switches back to the binary 0 level for the last bit.

The bipolar method currently utilized by most all packet radio stations and supported by all the manufactured TNCs is known as *NRZ Inverted (NRZI)* or *NRZ Space (NRZ-S)*. IN NRZI, a binary 0 causes a switch (or transition) between signal levels while binary 1 remains at the current level. The signal levels will no longer be referred to as "binary 1 level" and "binary 0 level" because these terms no longer have meaning in this encoding technique. Instead, the two signal levels are called 'mark' and 'space' levels. The same binary sequence, 1001110, is sent starting at the mark level, switches to the space level for the next bit, switches to the mark level for the next bit, remains at the same level for the next three bits, and then switches to the space level for the last bit. (See Fig. 4-7.)

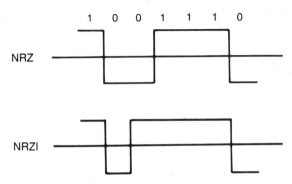

Fig. 4-7. A diagram showing the differences between the NRZ and NRZI encoding techniques.

Other forms of bipolar keying encoding techniques are *NRZ-Mark (NRZ-M)*; the opposite of NRZ-S, *Pulse Position Modulation (PPM), Pulse Duration Modulation (PDM)*, and Manchester I and II. For more information on these and other encoding techniques, see pages 19-39 in the 1986 *ARRL Handbook*.

Data Link Layer

The data link layer is also standardized in amateur packet radio. The AX.25 protocol is the most common, and is supported by most all commercial TNCs. The V-1 (VADCG), and V-2 are also data link layer protocols which differ from AX.25 in many respects. All three protocols are based on the HDLC ISO standard. Therefore, let's start off with a look at HDLC.

HDLC

HDLC is an acronym for High-level Data Link Control. The HDLC is the data link layer (level 2) of X.25 and is defined in the following ISO standards: ISO 3309, ISO/DIS 4335, ISO/DIS 6156, and ISO/DIS 6259. HDLC is responsible for delivering error-free data through the network. HDLC also isolates the upper levels from the physical layer. Data is broken up into blocks (frames) for transmission. The user data, the actual data being sent through the network by the users, is called data or information.

HDLC consists of three sublayers:

- transparency of the bit stream
- frame format
- cooperation between stations

Before going any further with the discussion of HDLC, let's digress for a moment and explain the differences between COPs and BOPs.

COP is an acronym for *Character Oriented Protocol*. Examples of COPs include ANSI X3.28 and IBM Binary Synchronous Communication. In a COP, the data being sent must be represented as characters of specified length, usually seven or eight bits (one byte). Thus, there is a limit to he type of information that can be transmitted. All transmission lengths must be a multiple of the specified character length. COPs are fine as long as only text is being transmitted. But packet radio is designed to be capable of sending any type of digital data including characters of different lengths, graphics, and special formats. In packet radio, it is also necessary to condense the size of the transmissions as much as possible.

This is where *Bit Oriented Protocols (BOPs)* come in. A good example of a BOP is HDLC along with the amateur packet radio data link layer protocols. BOPs allow for the transmission of any format of digital data. If control information is only 3 or 4 bits long, it only consumes 3 or 4 bits, not a full 7 or 8 as in COPs. Now, back to HDLC.

The first sub-layer of HDLC, transparency of the bit stream, means that all data being transmitted must look the same. No special length bits or signaling elements can be used. All data should pass through the physical layer without any altering or processing. HDLC must be independent of the data sent; however, HDLC must also delimit (mark the beginning and end) frames.

This is accomplished by the use of flags. A flag is a special binary sequence found only at the beginning and end of a frame. The flag used in HDLC is 01111110. This flag must not appear anywhere else in the frame except at the very beginning and very end.

To keep flags out of the user data, the data is examined, and a 0 is inserted after every five 1 bits found in a row. This is called *bit stuffing*. The receiving stations correct for this by removing the bit after a sequence of five 1s if it is a 0; if the bit is a 1, it does not, as the sequence of 1s is part of a flag.

In the second sub-layer, the frame format, all data is segmented and sent in frames which are delimited by flags. The components of the frame between the two flags consist of addresses, control information, data, and the *Frame Check Sequence (FCS)*. Figure 4-8 shows the HDLC frames. The following section describes the various frame components.

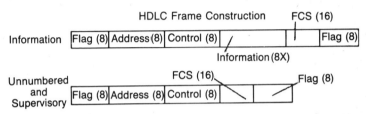

Fig. 4-8. The HDLC frames. The numbers in parentheses indicate the total number of bits in that particular field. The information field is usually a multiple of eight bits.

The first frame component following the initial flag is the *address*. The address is HDLC consists of the address of the originating station and the destination station. The addresses are usually numeric; however, AX.25 utilizes callsigns as addresses and includes digipeaters in the address section.

The next section (or field) in the HDLC frame is control information. Control information can consist of several things depending on the type of frame. There are three types of frames defined under HDLC: information, supervisory, and unnumbered. The control field consists of 8 bits.

Information frames are used for data transfer (they carry user information). Bit 1 of the control field of an information frame consists of a binary 0, bits 2 through 4 represent the transmitting station sequence number (often called transmit count), bit 5 is the poll/final bit, and bits 6 to 8 represent the receiving station sequence number (receive count).

Supervisory frames are used to control the flow of data. Bit 1 of the control field of a supervisory frame consists of a binary 1, bit 2 is a binary 0, bits 3 and 4 represent the supervisory frame type indicator. Bit 5 is the poll/final bit, and bits 6 to 8 represent the receive count.

Unnumbered frames are used to control the link. Bits 1 and 2 of the control field of an unnumbered frame consists of binary 1s, bits 3 and 4 are modifier bits, bit 5 is the poll/final bit, and bits 6 to 8 are also modifier bits.

Each station involved in a link maintains counters for the number of information frames sent and received. These counters are known as the *sequence number* or *count*. The counters are sent in the control field of each information frame, and are used to check the sequence of received frames and to acknowledge the reception of frames.

The poll/final bit is used to indicate that the receiving station must acknowledge the frame.

The last component of an HDLC frame next to the final flag is the FCS. The FCS is a *cyclic redundancy check (CRC)* performed on a frame. The FCS is an error detection scheme in which a check character is generated by dividing the entire numeric binary value of a block of data by a generator polynomial. The FCS value is sent along with the data, and the destination station, the FCS is recomputed from the received data. If the received FCS matches the one generated from the received data, the data is considered error-free.

The FCS is computed starting with the first bit after the opening flag and ending with the last bit preceding the FCS. For more detailed information on the FCS methods, see ISO standard 3309.

This completes the discussion of the components of a HDLC frame.

The last sub-layer, cooperation between stations, is handled by special frames which are recognized by HDLC as commands and responses. These command and response frames are used to establish connections, acknowledge receipt of frames, handle disconnections, and other tasks. Specific information about the different commands and responses used by HDLC to manage the link layer are not given here. The commands and responses utilized in amateur packet radio are discussed under the appropriate protocols.

AX.25

As you may recall from Chapter 2, the AX.25 protocol was originally developed by AMRAD, and after some minor modifications, AX.25 became the standard level 2 protocol in amateur packet radio. The AX.25 protocol is very similar to the level 2 protocol of the X.25 standard; thus, the name "AX.25" ("A" is for Amateur). This version of AX.25 filtered around for awhile, and minor incompatibilities existed between various implementations. In 1984, the ARRL Ad Hoc Committee on Amateur Digital Communications finished a revised version of the AX.25 standard.

The ARRL sponsored version of AX.25 is known as AX.25 Version 2 to differentiate their version from earlier versions of the same protocol. The earlier AX.25 protocol is now known as AX.25 Version 1. The most popular, and thus standard, AX.25 Version 1 protocol is the one developed by TAPR. While there are some incompatibilities between the two versions of AX.25, they are very similar.

AX.25 follows the same frame format as HDLC. The main differences are that the address field has been extended to allow for amateur radio callsigns as addresses and that unnumbered information frames (unnumbered frames containing user data) may be transmitted.

This is a brief description of the various components of an AX.25 frame (See Fig. 4-9.)

AX.25 Frame Construction

	Flag	Address	Control	PID	Information	FCS	Flag
Information and Unnumbered Info	8	112→560	8	8	≤2048	16	8

	Flag	Address	Control	FCS	Flag
Supervisory and Unnumbered	8	112→560	8	16	8

Fig. 4-9. The AX.25 frames. The numbers in each field indicate the number of bits contained in that field. The address field can contain a minimum of 112 bits (two callsigns at 56 bits each) and up to 560 bits (two callsigns plus eight digipeaters). The largest information field is 2048 bits.

The flag is identical in function and design to that used in HDLC.

The address field consists of a minimum of one amateur radio callsign belonging to the sending station in an unnumbered information frame (the destination address is set to a dummy address). In most cases, the destination station's callsign is included. Up to eight digipeater callsigns may also be added; thus, the maximum number of callsign addresses in the AX.25 address field is 10.

Each callsign consumes seven groups of eight bits (eight bits = one byte = one character; also called an *octet*). Callsigns are made up of uppercase ASCII characters and numbers. The first six characters are allotted for the actual callsign; if the callsign is less than six characters in length, spaces are added to the end. The seventh character is the *Sub-Station IDentifier or Secondary Station IDentifier (SSID)*. The SSID ranges from 0 to 15. Only four bits of the eight available for the SSID are consumed. The first and last bits are set to 0 and the remaining two are reserved for future use.

The SSID of a digipeater carries additional information. The last bit of the SSID is set to 1 once the frame has been digipeated by that station. This is necessary to avoid a digipeater repeating a frame twice.

The control field is made up of one octet. It is used to identify the type of frame (information, unnumbered, unnumbered information, and supervisory). It also contains the frame count numbers used for acknowledgements and special signals for establishing and maintaining connections (commands and responses).

There is a *Protocol IDentifier* field *(PID)* included with frames containing information. It is used to identify what kind of network layer protocol, if any, is being used.

The information field contains the user data being transmitted. The field is usually divided into a multiple of octets; the maximum being 256.

The FCS is the same as that used in HDLC.

Each transmission is usually preceded by a series of sixteen alternating bits to allow the receiving station time to synchronize onto the signal.

AX.25 is actually a subset of HDLC standard as it does not implement the full range of features defined in HDLC. AX.25 is based on the LAPB subset of HDLC. *LAPB* is an acronym for *Link Access Procedure Balanced*.

In a "normal" HDLC network, there are usually several slave stations (user terminals) linked to a central controller (host system; also called primary or master station). The master station usually contains more "intelligence" than the slave stations and is better able to manage the link. This configuration, with one intelligent master station linked to a less intelligent slave station, is called *unbalanced* because the stations do not possess equal capabilities.

In amateur packet radio, we want each station to be equal to any other station in terms of capability. Thus, no master or host station is used. The links between stations are in a balanced configuration. This type of station arrangement is supported by LAPB under HDLC.

Let's take a look at the actual frames used by AX.25 to establish connections, acknowledge frames, and disconnects.

Unnumbered frame control fields are either commands or responses. Unnumbered frames are used to handle all communications between stations when no connection has been established. There are six different unnumbered frames defined in AX.25.

SABM. The first unnumbered frame is a command called SABM. *SABM* is an acronym for *Set Asynchronous Balanced Mode*, and is used to send a connect request to another station. The control field in this frame contains a special sequence of bits which identify it as a SABM frame. The SABM command is used to place two stations in asynchronous balanced mode (meaning they are connected).

DISC. The next unnumbered frame is also a command. It is known as a DISC frame. The *DISC* frame is used to send a *DISConnect* request to another station, terminating the connection between the two stations.

DM. The *Disconnected Mode (DM)* unnumbered frame is a response. It is sent whenever a station receives any frame other than a SABM while disconnected. It is also sometimes sent in response to a SABM frame to indicate the station is not available for connection.

UA. The *Unnumbered Acknowledge (UA)* unnumbered frame is a response. It is sent as an ACK for unnumbered frame commands. A received command is not executed until a UA frame has been sent.

FRMR. The *FRaMe Reject (FRMR)* unnumbered frame is a response. It is sent when a frame is received that is not processable by the station (the frame is not garbled, but is not a valid frame). This response is usually sent when a frame clears the FCS check, but is not recognized by the station's protocol. Some situations which might cause this include the reception of a command or response not defined in the protocol or an information frame whose information field exceeds the maximum allowable length.

UI. The *Unnumbered Information (UI)* frame is an addition to the X.25 protocol included in AX.25. The UI frame allows for the transmission of an information field without first having established a connection. These frames are not acknowledged.

Once a connection is established utilizing the above commands and responses, it is up to AX.25 to handle the error-free transfer of information between the two stations. This is accomplished through the use of supervisory frames. There are three supervisory frames defined in AX.25.

RR. *RR* is an acronym for *Receive Ready*. RR is a response used to indicate that the sending station is able to receive information frames, to acknowledge the reception of information frames, and to clear a RNR response (see next paragraph) previously set by the station. The RR frame acknowledges the reception of information frames by including the receiver count indicating what frames have been correctly received. The other station can examine the count and update the next frame to be sent according to what frames can now be forgotten.

RNR. *RNR* is an acronym for *Receive Not Ready*. The RNR is a response used to indicate that the sending station is temporarily busy and is not able to accept any more information frames. This might occur when the receive station's receive buffer is full; it sends an RNR to the other station telling it to hold any information frames until the receiving station is able to receive more frames. The RNR condition can be cleared by sending a UA, RR, REJ, or SABM frame.

REJ. *REJ* is an abbreviation for *REJect*. The REJ frame is a response that is sent to request the retransmission of information frames that received out of sequence. The frame(s) to be retransmitted are indicated by the receive count in the control field of the frame. The REJ condition is cleared by the proper reception of the requested frames.

AX.25 Protocol Trace

Let's trace the steps taken by AX.25 when connecting and disconnecting from another station. When a user types C KR3T at the command prompt and presses the RETURN key, the TNC prepares a SABM frame containing the user's call and KR3T in the address field. The TNC then checks to see if the channel is available (CSMA). If so, the SABM frame is transmitted.

KR3T receives the frame and finds its call in the destination address. If CONOK is ON and the TNC is free to establish a connection, KR3T sends a UA frame to the originating station and enters the connected mode. If CONOK was OFF, KR3T would send a DM frame to the originating station.

Assuming KR3T is available and the originating station has received the UA frame from KR3T, it also enters the connected mode. Information frames are now used to transfer information between the two stations. The receive and transmit counters are kept current by the two stations.

The RR frame is used to acknowledge the reception of information frames. If a frame is received out of sequence by one of the stations, it sends a REJ to the other station and the frame is retransmitted.

When one of the stations wants to end the connection, he returns to the command mode and enters D at the command prompt. His TNC then sends a DISC frame to the other station. The other station sends a UA frame acknowledging the disconnect request and then enters the disconnected mode. Once his station receives the UA frame, it also enters the disconnected mode.

AX.25 Conclusion

This concludes the section on the AX.25 link layer protocol. You may have noticed a few points about AX.25 which do not seem to fit in with the earlier description of the data link layer given earlier in the chapter. These discrepancies include the addition of digipeaters and the use of end-to-end acknowledgements. Digipeaters seem to fall under the control of the network layer protocol, not the data link layer. In many respects this is true.

However, keep in mind that digipeaters are not full-fledged level 3 network nodes by any means. They are a simple kludge added to allow for rudimentary networking. The user must select the digipeaters used; the TNC does not contain automatic routing tables or other means of independently selecting digipeater routes. Most likely, digipeaters will eventually fade out as more advanced network nodes and level 3 protocols emerge.

As for the end-to-end acknowledgements, in the OSI/RM these would seem to be the responsibility of the transport layer (level 4). The transport layer is responsible for the proper reception of frames from one station to the other through the network. However, in

AX.25 there is no network layer protocol yet implemented, so the point-to-point acknowledgements of the data link layer were simply extended over the digipeater path. If the digipeaters are eliminated from the path, the acknowledgement returns to a point-to-point ACK as used by the data link layer. Once network nodes are implemented, they will use point-to-point acknowledgement between nodes and a transport layer acknowledgement between the two end stations.

For more detailed information on the AX.25 protocol, I recommend the following sources:

- *The TAPR TNC-1 System Manual*
- "AX.25 Amateur Packet-Radio Link-Layer Protocol" from the ARRL
- "Proceedings of the Second ARRL Amateur Radio Computer Networking Conference" from the ARRL

V-1

The next packet protocol we will examine is V-1. The V-1 protocol is more commonly known as the VADCG protocol. It was developed by Doug Lockhart VE7APU in late 1979. In the summer of 1979, Doug was working on a protocol for the VADCG TNC which was designed for use with a master station and multiple TNCs. The TNCs would connect to the station node where they would be assigned numeric addresses and all communications between users would take place through the station node.

A packet group in Hamilton, Ontario was working on developing TNCs, and they wanted a protocol that allowed the TNCs to be connected to each other directly rather than going through a station node. They asked Doug if he could provide them with such a protocol so that they could concentrate on experimenting with the TNCs rather than having to set up a station node first.

Doug complied with their request by modifying the protocol he was using by eliminating the dynamic addressing and other facilities used by the station node. The end result was a simple protocol which allowed TNCs to connect to each other directly; a kind of kludge to meet a specific request.

Well, this temporary experimental protocol grew in popularity because it was distributed with VADCG TNC kits and eventually spread to the United States. This protocol is commonly called the

VADCG protocol. Now that Doug has completed work on a second VADCG protocol, this original VADCG protocol has been named V-1.

The V-1 protocol achieved widespread use in the United States due to the fact that it was the only protocol around that actually worked at the time. Doug never intended for it to become a standard, but with nothing else around it became a *de facto* standard. The VADCG protocol in use in the United States was not exactly identical to the one that Doug developed for the Hamilton packet operators.

The VADCG protocol as it was first used in Canada allowed for up to 254 numeric addresses. Each station would have one numeric address assigned beforehand. When Hank Magnuski KA6M put the United States' first all-digital simplex packet repeater (a digipeater) up on 10 December 1980 in California, he used a modified version of the VADCG protocol in order to allow for digipeating. Hank used a portion of the address space to support digipeater control, thus reducing the total number of user addresses from 254 to 32.

This modified version of VADCG protocol used in California became the standard in the United States. Doug was approached several times about making the VADCG protocol an international standard. However, he consistently resisted the idea as he still saw the VADCG protocol as a temporary testing protocol that was too limited for widespread use. These limitations would later cause packet groups in the United States to develop alternate data link level protocols such as AX.25.

The VADCG protocol had only two commands: connect and disconnect. To connect with another station, the user would type its callsign and a CTRL-X. A connect request frame would be sent containing the other station's call, the user's call, and the user's VADCG numeric address. The other station would respond with an acknowledgement containing this VADCG numeric address. From then on, all information frames would be sent with only the VADCG addresses. To disconnect, the user would type a CTRL-Y and a disconnect request would be sent. Once the disconnect request was acknowledged, the connection was terminated.

The VADCG protocol is still in limited use today in Canada and Australia. The VADCG protocol is very similar to HDLC as numeric addresses are used. The VADCG protocol will not be explained in detail, as it is no longer in widespread use and is of limited use to most packet operators.

For more detailed information on the VADCG V-1 protocol, I recommended the following references:

- *The TAPR TNC-1 System Manual*
- Early issues of *The Packet* Newsletter from VADCG

V-2

Doug Lockhart has continued with the development of link level protocols as he does not feel that AX.25 is a viable link level protocol for the large scale development of packet radio because of various technical reasons. The next VADCG protocol he developed is known as V-2. It offers many improvements over the V-1 protocol.

Both full and half duplex links are allowed. Multiple links are supported along with some multiple protocol support. The number of numeric addresses has been significantly increased, and no coordination of the address is needed beforehand.

When the specifications of V-2 were published, TAPR compared them with the AX.25 protocol and found some notable variations, but not enough of a difference to necessitate changing the current link level protocol. Some of the differences between V-2 and AX.25 are that V-2 offers reduced address space, differentiation between user names (callsigns) and node addresses (numeric identifier), and no "network level" functions.

V-2 is similar to SLDC, HDLC, AX.25, and V-1. V-2 is designed to be implemented along with upper level protocols to establish a complete network. For more information on the V-2 protocol, I recommend the following references:

- "Proceedings of the Third ARRL Amateur Computer Networking Conference" from the ARRL
- *The Packet* Newsletter, Issue 9: October 1984, from the VADCG

Other Level 2 Protocols

There have also been several other data link layer protocols in use in amateur packet radio over the years. Two Canadian groups, other than the VADCG, developed viable packet systems with their own protocols. TAPR developed their own link level protocol known as TAPR/DA prior to the adoption of AX.25.

One of the Canadian groups based in Montreal was using a combination COP/BOP protocol using ASCII characters as frame delimiters. Their system was operational in 1978 running at 4800 baud AFSK on 220 MHz.

A second Canadian group based in Ottawa was using a polling protocol. They first developed the concept of a digipeater. Their system was running at 9600 baud FSK in 1980.

The order of the very active early Canadian groups is Montreal in 1978, Vancouver in 1979, and Ottawa in 1980.

For more information on the additional Canadian systems, I recommend the following references:

Montreal.

- *Packet Radio* by Rouleau VE2PY and Hodgson VE2BEN from TAB Books (#1345)
- "The Packet Radio Revolution" by Rouleau VE2PY from December 1980 *73 Magazine*
- "An Introduction to Packet Radio" by Hodgson VE2BEN from June 1979 *Ham Radio Magazine*

Ottawa.

- Early issues of *The Packet* Newsletter from the VADCG

TAPR had a data link level protocol up and running shortly after their founding. It is known as TAPR/DA for TAPR/Dynamic Addressing. It was being implemented before the 1982 AMSAT/AMRAD protocol conference and was one of the protocols considered. However, it was passed over in favor of AX.25 during the conference.

NETWORK LAYER

Current protocol work in the United States is being directed towards level 3 network layer protocols. There are several systems under development today. Most of the protocols are based in the level 3 of X.25.

One system is being experimented with by the RATS group in New Jersey. Another is being developed by the *Florida Amateur Digital Communications Group (FADCA)* group in Florida. (See Fig. 4-10.) At their current stage, these systems simply acknowledge

78 Networking and Protocols

Fig. 4-10. A Xerox 820 computer configured for use as a level 3 network node. This system was exhibited at the Fifth ARRL Amateur Radio Computer Networking Conference by FADCA and is called the "Gator 2 Switch." (Photo courtesy of FADCA.)

packets being repeated through them (they replace the end-to-end ACK of AX.25 with a point-to-point ACK). In the near future, they should contain routing tables that will automatically select the best route for a packet to reach its destination, transparently to the user.

Addressing for the network nodes is a concern. The RATS group is basing their address scheme on the X.121 standard which defines numeric addresses for the nodes. The FADCA group favors alphabetic designators. However, this is a minor problem which can be easily resolved.

Another more recent networking system is NET/ROM developed by Ron Raikes WA8DED and Mike Busch W6IXU. This protocol simply replaces the ROM chip in any TNC-2 compatible TNC and turns the TNC into a network node. Users can connect to other NET/ROM nodes as well as conventional stations with the benefit of point-to-point ACKs through the NET/ROM nodes. It also maintains routing tables to other local NET/ROM nodes so it is not necessary to specify a path between nodes. There are now numerous NET/ROM nodes operating all over the country. Operational information on NET/ROM is included in Chapter 6.

TAPR's latest project is a *Network Node Controller (NNC)*. The NNC is a combination hardware and software device designed to serve as a network node in an amateur packet network. It is actually a small, but powerful, computer system complete with microprocessor, memory, and external storage. There are four HDLC ports with very flexible characteristics.

Two ports might be configured for Bell 202 use on VHF, a third for 9600 baud on UHF, and the fourth for Bell 103 on HF. Thus, a user may connect to the NNC on one port and route his packets out one of the other ports (this may be handled automatically by the NNC's software). The NNC will acknowledge frames as they are received (a point-to-point acknowledgement).

The network layer (level 3) is usually divided into two distinct sub-levels: 3A and 3B. Level 3A is controlled by the *intranet* protocol and 3B is controlled by the *internet* protocol. The intranet protocol deals with communications around a single network node and user stations. The internet protocol deals with communication between network nodes.

Communications between individual users and their network node (intranet) will most probably be accomplished through *virtual circuits*. A virtual circuit is a method of connecting stations in which an abbreviated address field is used once a connection is established. The stations must be connected before communications can begin, and after the connection is established, the addressing information contained in each transmission is decreased. However, the lack of complete addressing information forces each packet to take the same path through the network.

Communications on the internet are still under debate. They may be virtual circuits or possibly datagrams. A *datagram* is a method of connecting two stations in which each packet sent over the network contains complete addressing information. The advantage of datagrams is that they may be dynamically routed through the network (i.e., the path of the connection may change) because they contain complete addressing information. The advantage of virtual circuits is that once the connection is established, the amount of packet space consumed by the address field can be significantly reduced.

Virtual circuits require more reliable, intelligent network nodes to remember the path of the connection. Datagrams may reach their destination despite the failure of one or more network nodes by being dynamically routed around the nonfunctional nodes.

Once network nodes are fully implemented, a user from one area of the country can access a node and connect to any other station that the network reaches (similar to the telephone system). More information on possible network architectures is given in Chapter 8.

For more information on level 3 networking see the "Proceedings of the Third ARRL Amateur Radio Computer Networking Conference" from the ARRL and the 1985 or 1986 *ARRL Handbook*, Chapter 19, from the ARRL.

Transport and Session Layers

Levels 4 and 5 of the OSI/RM have also been under development for use in amateur packet radio. A popular contender for the level 4 transport layer is *Transmission Control Protocol (TCP)*. It is a complex protocol designed for dealing with inadequate or unreliable lower-layer protocols. The session layer (level 5) is also handled by TCP. For more information on TCP, I recommend *Computer Networks* by Tanenbaum from Prentice-Hall.

Phil Karn KA9Q has developed a *Transmission Control Protocol/Internet Protocol (TCP/IP)* program for the IBM-PC. This program (NET.EXE) allows for advanced networking in amateur packet radio, although the TCP/IP system is still very experimental in nature. NET.EXE, its related programs, a file and message transfer system, and documentation can be downloaded from CompuServe's HamNet.

Presentation Layer

The level 6 presentation layer has also been given some thought. The *North American Presentation Level Protocol Syntax (NAPLPS)* graphics protocol may be a means for sending graphics via packet radio. Other contenders include *File Transfer Protocol (FTP)* and *Simple Mail Transfer Program (SMTP)*.

CONCLUSION

This chapter has covered most aspects of amateur packet radio networking and protocols. If you have an interest in learning more about networking and protocols, take a look at a few of the networking books listed in the Bibliography. If you have had some difficulty understanding the various concepts presented here, do not despair. In-depth knowledge of networking and protocols is not a prerequisite to operating and enjoying packet radio.

The next chapter describes the details involved in setting up your own packet radio station.

Chapter 5

Setting Up an Amateur Packet Radio Station

Let's take a break from the more technical aspects of packet radio to discuss something that all users, from novice to expert, have to do to join in on the fun of packet radio—set up a station. You have already been introduced to the necessary components in the earlier chapters, so in this chapter, we'll take a look at TNCs, terminals, modems, and radios from an operational and pragmatic, not technical, perspective. Several examples of packet stations are given along with suggestions to help you maximize your station's capabilities based on your particular situation. Station arrangement, RFI protection, antennas, power requirements, and portable operation are also discussed.

Obviously, everyone has a different set of objectives when setting up a packet station. For some, packet takes a back seat to other methods of communication. Others see packet as their primary mode. Some want to operate HF, while others prefer VHF operation. There is no ideal station for everyone, so my advice to you is to read through the chapter and make note of anything you feel might be of use when setting up your station or modifying your present station.

The first major decision that must be made when assembling a packet station is which terminal and TNC to use. They are listed together because they must work together and be fully compatible. The next section describes several types of terminals and lists their features so you can compare them and see which is best for you.

In some cases the terminal and TNC are combined into a single unit. This is often true of a machine-specific TNC (a TNC designed to work with only certain computers or terminals) and software-based TNCs where a limited number of computer systems are supported. While this can be very convenient and greatly reduces the decision-making required when selecting a terminal and TNC, keep in mind the inherent lack of options. A combination terminal/TNC may have a predefined user interface, computer system requirements, and other non-variable features. These options are usually much more flexible when utilizing a universal TNC and a universal terminal. However, one great advantage that combination terminal/TNCs have is that they are designed to work together. The TNC may utilize features peculiar to only that terminal. The user interface, if well thought out, can be enhanced by taking advantage of those features.

SELECTING A TERMINAL

Most TNCs are designed to communicate with a wide variety of terminals through a standard I/O interface such as RS-232, TTL, or current loop. While almost any terminal will work with TNCs, there are many options that need to be considered when selecting a terminal in order to achieve maximum flexibility from your packet station. Terminals come in many different configurations and understanding the features available from each can help you pick a terminal that best suits your operating style.

There are two main types of terminals: *dedicated terminals* and *terminal emulators*. A dedicated terminal is a device whose sole purpose is to convert digital codes into recognizable symbols. A terminal emulator is a software program which allows a computer system to act like (emulate) a dedicated terminal. In addition to the distinction between dedicated terminals and terminal emulators, terminals can also be classified as either dumb or smart. We'll look first at the difference between dedicated terminals and terminal emulators, and then examine smart and dumb terminals.

A dedicated terminal is designed to function only as a terminal, to convert digital signals to symbols and symbols to digital signals. Almost all terminals receive digital ASCII or Baudot signals through an I/O port and display them in their alphabetic equivalent. Information is typed on an alphabetic keyboard, and the information's ASCII or Baudot equivalent signal is sent out the I/O port. These terminals are called either ASCII terminals or Baudot terminals, depending on what code the terminal uses. ASCII terminals are the most common type used in packet radio. (Other types of terminals, such as graphics terminals, are not presently in wide use in packet radio.)

A hardcopy terminal generates its output on paper utilizing some form of printer. The most common form of hardcopy terminal is the mechanical teleprinter such as the Model 33 ASR. Mechanical teleprinters are not recommended for serious packet operation due to their slow speed, maintenance requirements, and noise. Modern hardcopy terminals usually incorporate a great deal of electronics and generate their output using high speed dot-matrix, thermal, or ink jet printers. Modern hardcopy terminals operate at a speed sufficient for serious packet operation, but lack many of the features available in video display terminals (not to mention the problem of what to do with all the paper these terminals crank out).

Video or CRT dedicated terminals are very useful devices which can have many features. Some of the terminals are as electronically complex as microcomputers. These terminals usually feature memory, full keyboards, and ergonomic design, and offer expansion such as printers and additional features. Video terminals are usually smart terminals.

Rather than use a dedicated terminal, most packet operators utilize a terminal emulator for communicating with their TNCs. A terminal emulator is a software program which runs on a microcomputer and allows the microcomputer to function as a terminal. The terminal emulator concept is very popular for several reasons. It is cost effective because there is no additional equipment to buy beyond the computer system, and the computer may be used for other purposes. A computer can be used for a wide variety of functions by changing software while a dedicated terminal can only be used as a terminal.

Computer-based terminals also allow for much greater flexibility in handling the information obtained through the terminal. With a dedicated terminal, received information can only be printed out (or with some CRT terminals, temporarily stored). With a computer-based terminal emulator, the information can be permanently saved to disk or cassette, formatted and printed using word processing software, or processed using database or spreadsheet software.

Terminal emulator software comes with a wide variety of capabilities. Some simply allow the computer to serve as a dumb terminal. Other, more complicated software, allows the microcomputer to function as a sophisticated smart terminal.

Some terminal emulation software is written especially for packet operation. This software is usually designed for use with a specific TNC and contains features which enhance its use. However, it is not necessary to have a terminal program exclusively for packet operation.

Most ASCII telecommunications software packages will work. These are the same programs used with telephone-based modems to access remote computer systems. There are a great number of terminal programs with an equally great diversity in features and capabilities. Generally, microcomputer systems with communications capabilities have terminal software available for them.

While terminal software can cost up to several hundred dollars, it is not necessary to spend vast sums of money. There are a great variety of terminal programs available in the public domain which may be obtained at little or no cost. Often, operating system master disks include some sort of telecommunications capability, especially if the computer comes standard with an RS-232 port. If you're not sure what kind of program is best, or where to look for one for your particular computer, check with someone who is using the same type of computer you do. If you don't know of anyone, try to locate a users group in your area; a good place to start is where you bought your computer. Be sure to look at a few back issues of computer magazines for ideas; check to see if there is one written specifically for your computer system.

I've been using the terms smart and dumb terminals throughout the preceding section without much explanation. Let's discuss the differences between smart and dumb terminals. Keep in mind that the "terminal" can be either a dedicated terminal or a terminal emulator.

A dumb terminal has only the most basic of capabilities necessary for communications. It features very simple send and receive functions, most often just displaying received characters and transmitting exactly what was typed. Dumb terminals do not support expansion, nor do they offer much flexibility for modifying the communications characteristics. A good example of a dumb dedicated terminal is a mechanical teletype. A dumb terminal emulator is a very simple terminal program.

A smart terminal possesses advanced capabilities to enhance communications. Smart terminals are usually dedicated CRT terminals or sophisticated terminal programs. There are many features available in smart terminals. The most useful are listed below.

Xon and Xoff capability is a very useful, almost mandatory, feature for packet operation. The Xoff signal, usually a CTRL-S, tells the terminal to stop sending information. An Xon signal, usually a CTRL-Q, tells the terminal to resume sending the information.

This feature allows an attached device to automatically turn the flow of data from the terminal to the attached device on and off so that the device is not overloaded. If the device supports Xon/Xoff, then the terminal can turn the flow of data from the device on and off so that the terminal is not overloaded. For example, Xon/Xoff is useful when a TNC is transmitting at a much lower baud rate than is used between the TNC and the terminal. Assume the TNC's modem is transmitting at 300 baud and the terminal is sending the TNC data at 9600 baud. The terminal is sending information to the TNC faster than the TNC can send it out, so the TNC buffers the information build-up in its memory. When the memory is almost full, the TNC sends the terminal an Xoff character to stop the flow of data. When the buffer space is again at an acceptable level, the TNC sends an Xon character to resume the flow of data.

Other control codes besides Xon and Xoff are utilized heavily in digital communications to control the receiving terminal. Control codes such as clear screen, bell, and linefeed are commonly used in packet. A smart terminal should allow for changing or filtering of control codes. For example, while many control codes are standardized (i.e., Linefeed = CTRL-J and Bell = CTRL-G), many are not. One terminal's clear screen might be another's delete transmit buffer. One of my terminal programs uses CTRL-Z as the clear screen command which also happens to be the W0RLI Mailbox's end of file character. Because of this, whenever I received an end of file my screen would go blank. By filtering out the CTRL-Z character before it was processed by the program, the screen would not clear at the end of a message, and I was able to read the messages in their entirety without the screen suddenly clearing.

The ability to download a file is extremely useful in all digital communications, including packet, because it gives you the ability to save incoming data for later viewing or processing. On a dedicated terminal, the data is usually saved in volatile memory. On a computer system with a terminal emulator, the data may be put in volatile memory and then saved to disk or cassette. The computer has the distinct advantage in this area due to its long term storage and processing capabilities.

Terminals differ on downloading procedure, but most allow the user to open a buffer with a special keyboard sequence and close the buffer with another sequence after the desired data has been received. The buffer may then be cleared, displayed, printed, or, with computer and certain dedicated terminals, saved on disk or cassette. These functions may be handled through special commands or through menus. A nice feature to have in a terminal emulator is the ability to turn menus off once you have gotten used to the structure and commands.

Uploading is another useful feature that smart terminals offer. Uploading allows for the sending of prepared data. The data may be saved on disk or in memory. In most systems, you specify which file or buffer is to be sent, and the data is automatically sent. This feature saves time and allows the same information to be sent to multiple destinations with a minimum of effort.

Echoing allows data to be sent to another device at the same time it is being sent or received by the terminal. In most cases, the other device is a printer. With echoing you can get a hardcopy of data without downloading it into memory. This comes in very handy when downloading a file that exceeds the available memory on dedicated terminals and non-disk computers.

Keep in mind that many terminal systems have more features than I have listed here; these are the basics that you should look for in a smart terminal for use with packet. The dividing line between smart and dumb terminals is not black and white, so more attention should be given to a system's features rather than its designation. Some terminals are smarter than others, but with increased capability usually comes increased complexity. The bottom line is to find a terminal that has the features you need and feel comfortable using.

My recommendation as to the first terminal you should consider is a microcomputer system with a smart terminal emulator program. This provides maximum flexibility. You can retain information permanently, and with a CRT, you can control the amount of paper you generate. However, a computer system with disk drives, RS-232 port, and a printer (see Fig. 5-1) can be expensive. Be sure to take a look at the used computer market. If necessary, you can get by without a disk drive or a printer for awhile and add them later.

If you already own a computer, then it's time to look for some good terminal software. Check the computer to see if you need to add an RS-232 port for serial communications or a printer port. If you already have these things, then you're all set.

If you happen to share a computer with the rest of the family and cannot afford one exclusively for amateur radio use, do not despair. You can use the computer for occasional downloading. A cheap dumb terminal can be used for everyday monitoring and ragchewing. A computer is very useful in the shack, and you should seriously consider putting aside a little money towards one exclusively for amateur radio use.

If you don't have a computer and don't want one, a dedicated terminal can be used. If you decide on a hardcopy terminal, try to get one with as high a baud rate as possible. Twelve hundred baud

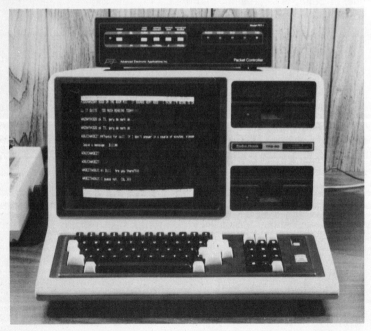

Fig. 5-1. A TRS-80 Model IV (running terminal emulation software) connected to an AEA PKT-1.

is the minimum limit for moderate use. Mechanical teletypes such as the Model 33 are just too slow for use as anything other than monitoring activity or infrequent operation.

Make every effort to get a terminal with an RS-232 port. Almost all TNCs use it for terminal communications. Besides, with an RS-232 port, the terminal can be connected to a multitude of other devices. Try to avoid current loop if at all possible. For more information on terminal devices along with information on interfacing teleprinters, see *TV Typewriter Cookbook* by Lancaster from Sams.

Many inexpensive home computers such as the Commodore 64 contain a serial I/O port. However, their I/O ports use TTL levels and are not usually compatible with RS-232 levels. Thus, it is necessary to convert the port to RS-232 levels (which is fairly easy to do) or to select a TNC which is compatible with the TTL levels.

Current loop is another form of digital I/O that can be used with some packet radio TNCs. It is not very convenient and performance suffers, but it is possible. Current loop systems are usually associated with mechanical teleprinters. They were very common in RTTY systems before the advent of solid state systems. There are many current loop systems around available at very low prices. Since most TNCs utilize RS-232 compatible I/O, an RS-232 to current loop adaptor must be attached to the TNC.

The TNC is then put into the loop along with the teleprinter and paper tape devices. Flow control is a common problem experienced when adapting an RS-232 device to a current loop system. If you are not interested in having a complete loop, but simply using the teleprinter as a terminal for the TNC, the TNC can be directly connected to the teleprinter. A special interface can be built which converts the RS-232 level signals to current pulses which trigger the teleprinter. Alternately, if an RS-232 card is available for the teleprinter, one may be installed and the TNC can be wired directly to it.

SELECTING A TNC

When selecting a TNC, several items should be considered. The TNC's command set, or user interface, controls your access to the TNC. Command sets vary from full English commands to short, sometimes cryptic, abbreviations. The TAPR command set has become the industry standard and most present-day TNCs utilize a similar command set. The advantage of an English syntax is the ease of remembering the commands. Abbreviations have the advantage of being faster to type once you have grown used to them. The TAPR command set (which will be covered in greater detail in the next chapter) allows users to type the full command (Connect) or an abbreviation (C). In some cases, more than one or two characters need to be typed to distinguish between similar commands (for example, MYC for MYCALL and MYV for MYVADCG).

Another area of possible concern when selecting a TNC is power consumption. This is especially important if you are planning portable operation. One way to lower the power consumption of a TNC is to replace NMOS chips with their less power-hungry CMOS equivalents, if they're available. Some TNC manufacturers offer low-power versions of their TNCs.

The TNC's buffer size can be important if you are using a slow terminal or doing extensive off-line monitoring. A slow terminal causes information to back up in the TNC and sufficient buffer size is necessary to hold the data until the terminal is ready to receive it. If the terminal is disconnected or otherwise taken off-line, the TNC can buffer received information until the terminal is back on-line. The amount of memory that comes with most TNCs is more than adequate in most cases. If you find that it is not, additional memory can usually be added on later.

The TNC is the most important piece of equipment in a packet station; thus, it is a good idea to pick your TNC first and arrange your station around it. Only compromise on the TNC if packet will not be one of your primary modes.

The modem is usually built into the TNC. If it's not, check the manufacturer's recommendation. Select the correct modem type for the band of operation: VHF-Bell 202 and HF-Bell 103. Some characteristics to look for in a modem for packet operation are a fast carrier detect and a non-PLL demodulator for HF operation. Status indicators for Transmit (TX or PTT) and *Data Carrier Detect (DCD)* are helpful. Be sure to check the interfacing data for both TNC to modem and modem to radio.

SELECTING A RADIO

A transceiver is the most convenient radio system for packet operation. The rig should meet the characteristics listed in the last chapter. One of the most important characteristics is the R/T and T/R times (the time it takes a rig to switch from receive to transmit

Selecting a Radio 91

and the time from transmit to receive). Try to get a rig with as low a value as possible for both. Once again, for VHF most any FM transceiver will work fine. On HF, a stable SSB transceiver is required. Also, a tuning indicator for the TNC is very useful. (See Fig. 5-2.)

The modem-radio interface is usually not very difficult. In most cases there are just two cables running from the modem to the rig: the mic cable and the phono or external speaker cable. There are usually no internal connections to the rig unless you are utilizing squelch detect or bypassing certain sections of the rig.

The carrier sense system does a good job of keeping a packet station from transmitting on top of another packet station, but it does not do a good job of keeping a packet station from transmitting on top of phone operators. If sharing a packet channel with voice users you should connect the squelch detect option. However, not all TNCs have squelch detect.

The squelch detect lead connects to the rig where the voltage varies according to the squelch condition. Most TNCs that implement this feature look for a voltage that goes from under 2 volts to over 3 volts when a signal is received. If your rig's voltages go in the opposite direction, an inverter is needed to get the proper voltage

Fig. 5-2. The radio equipment at KR3T. A Kenwood TS-430S is used on HF, an ICOM 271A is used for 2-meter base operation, and an IC-02AT is used for portable operation.

levels for the TNC. A good spot for tapping the voltage is the receive or busy indicator light on most rigs.

The microphone (mic) connector has three main connections: mic audio, PTT, and ground. The mic audio carries the AFSK from the modulator to the transmitter. The PTT trips the rig into the transmit mode. Be sure to check your rig's PTT characteristics to see if they're compatible with the modem's. The PTT line is usually a direct connection, but an interface may be required for some rigs. The ground shields the mic audio lead and should be at the same potential as the radio's ground.

The speaker connection runs to the modem's demodulator. The cable consists of an audio lead along with a ground. There are three possible taps for the audio: phono jack, external speaker jack, and discriminator. The phono jack just plugs into the headphone connector on the front of most rigs. There are usually two positions: all the way in for no internal speaker output and part of the way in for both phono plug and speaker output. The external speaker jack is usually located on the back of the rig, and when the plug is in, no audio goes to the internal speaker. Some TNCs do have an external speaker jack on them which allows monitoring of the received signal. When connecting to the discriminator, you get the signal before processing and possible distortion. This is an internal connection and in most cases need only be done when distortion is suspected in later stages.

The modem connection usually consists of a single plug, although many types of plugs are used. Some modems have separate plugs for audio connections so they have individual shielding to help reduce interference. Shielded cable is recommended for all connections, although ribbon cable can be used if there is no interference. Most modem plugs are designed to accept ribbon cable connectors.

If you have many mic connections, I do not recommend using separate mic plugs. They are expensive and constant insertion and removal will wear out the socket on the radio. The system I use is to wire a mic connector to a 5 pin DIN in-line socket. You can get away with 5 pins because the rest aren't needed for packet operation. A piece of shielded cable a foot or so long can be used to slightly lengthen the reach of the cable.

Then wire the modem's cable to a 5 pin DIN in-line plug. Make sure the pins are wired properly. Now insert the plug into the socket and you're all set.

All future connection to the same rig need only have a DIN plug, which costs about $1.50. Be sure to keep all cable runs as short as possible and the mic/audio lead shielded.

SAMPLE STATIONS

Individual packet stations vary greatly from one user to another. Each user's goals, budget, and priorities are different, which is reflected in the wide variety of set-ups around. It should be evident by now that despite the requirements of packet radio, there is quite a bit of latitude for the individual user to customize a station to meet his or her particular needs.

In this section, several stations are presented. They include a typical station, a budget station, a station with a shared computer, a station utilizing the current loop, and a portable station. Hopefully you will be able to find something useful in the examples that will aid you in setting up your own packet station.

Typical Station

The typical packet station arrangement used by an overwhelming majority of packet operators consists of a microcomputer with terminal emulator software, a RS-232 port, and disk-based storage; a TNC with an on-board modem; and a 2-meter FM transceiver. Optional equipment includes an external tuning indicator for HF operation, a separate HF modem, and a printer. Most of the characteristics of the components have been covered earlier, so I won't go over them again. The microcomputer can range from a Commodore 64 (see Fig. 5-3) or TRS-80 Color Computer to a TRS-80 Model IV or an Apple IIe (see Fig. 5-4), to IBM PC's and Compaqs. The terminal emulator software is usually very capable and offers a wide range of features. Most users have two or more terminal software packages that they are familiar with.

This is the arrangement that I recommend in most cases because it is widely used, well-tested, and extremely flexible. (See Fig. 5-5.) If one component needs replacing, it can be switched without altering the other equipment. For example, if you get a new computer, you can just unplug the TNC's RS-232 cable from the old computer and plug it into the new computer. The system can also be used for other modes such as Baudot and ASCII RTTY, CW, and AMTOR with the addition of proper software and an additional Terminal Unit. Also,

94 Setting Up an Amateur Packet Radio Station

Fig. 5-3. The Commodore 64 microcomputer.

Fig. 5-4. An Apple //e (running terminal emulation software), connected to a Kantronics KPC-2.

Fig. 5-5. A TRS-80 Model I (running terminal emulation software), connected to a Heathkit HD-4040. A Kantronics KPC-2 is set atop the HD-4040.

the computer can be used for other purposes such as logging and QSLing (acknowledging receipt of a contract).

The cost for this arrangement varies greatly. A low of $575 (a $300 computer system + a $125 TNC + a $150 radio) is easily realized, although used equipment may have to be considered. A high of $3950 ($3000 computer + $350 TNC + $600 radio) can be obtained by choosing fairly expensive equipment. Keep in mind that this system is usually implemented by users who already have a computer system available. If you don't have a computer yet, you should, and packet provides an excellent excuse to get one.

Budget Station

If you don't have much money to invest in packet radio, consider taking some ideas from the two budget systems I have arranged. One uses a software-based TNC and the other an inexpensive hardware system.

Preparing a packet station with a software-based TNC can be very inexpensive if you already have a compatible computer. In the case of the Richcraft software, that is a TRS-80 Model I, III, and IV. Model I and III disk-based systems are currently selling used for around $300. An external modem must be added and interfaced to the radio and the computer. One can be built for about $30 or you can search for a used one. The software costs about $29.

However, do not plan on keeping the software system for long. While it serves as an inexpensive means to access packet radio, it does not offer the features and flexibility found on universal hardware TNCs. But, if you happen to have a compatible computer or plan on buying one independently of the software, the software can significantly lower the cost of the packet system by eliminating the cost of a hardware TNC. A software-based packet system would cost about $359 ($300—computer, $30—modem, and $29—software). And if you already own the computer, the cost drops to $59.

An inexpensive hardware system is centered around the TNC. It is the most important part of the packet station and should receive the most consideration in a financially limited situation. The terminal and rig can be upgraded later, but with a flexible, capable TNC, almost any hardware configuration can be made to work.

I have divided the terminals into three levels based on cost. The lowest level consists of a mechanical hardcopy teleprinter with an

RS-232 port. Avoid this type of terminal if at all possible due to its slow speed! However, its price cannot be ignored. A mechanical hardcopy teleprinter can be had at a very low price or even free—I once received a perfectly good Model 33 ASR for free by just asking for it.

The middle level consists of a high speed hardcopy terminal. These units often have speeds in excess of 1200 baud and usually have a full implementation of the RS-232. But they generate a lot of paper and can be a little noisy, although those with thermal printers are usually very quiet. These terminals can be obtained at hamfests and computer shows for $50 to $100.

The high level of the budget terminal market consists of an inexpensive microcomputer or CRT terminal. An inexpensive microcomputer such as the Commodore 64 and the TRS-80 Color Computer sell for around $100 without accessories. By the time you add a cassette drive and software, the price is around $150. Don't forget that you'll also have to supply a monitor as these computers don't come with one. Check the used micro market for some good deals on complete systems.

A CRT terminal can usually be found surplus for around $100 to $150. Make sure it has an RS-232 port. There is nothing wrong with using a CRT terminal for casual packet operation, but a computer offers you more for the money.

The TNC is what you want to concentrate your available funds on. Check the used market. You might be surprised to find a TAPR TNC-1 or other early TNC at a great price. Another option to consider is purchasing a bare kit and building the unit yourself. The bare kit usually comes with a PC board and the EPROMs containing the system software. You supply the rest of the parts and the case. The cost is about $60 for the kit alone. The total price varies depending on how good you are at scrounging parts.

For the radio, just use a 2-meter HT or transceiver. If one is not available, get a simple crystal controlled channelized rig. Try to get a radio with as fast a T/R and R/T time as possible. Keep in mind that you might have to modify the rig to optimize it for packet radio. For HF operation, a clean, stable rig is needed along with an external tuning indicator.

Shared Terminal

If you have access to a microcomputer near your shack but are sharing it with the rest of the family or using it as a business computer, consider the following options. If the computer is located in a room other than your shack you can run an RS-232 cable from the TNC in your shack to the computer. As shown in Appendix B, it is not necessary to have all 25 pins connected, so you can get by with as few as three wires. As long as you don't need to tune the HF rig or rotate a beam from the shack, this system can work well. A VHF station can be set and left. You can use the computer for other tasks and then switch over to a telecommunications program when you wish to operate packet. An RS-232 switch may be needed to switch the computer's RS-232 port between the TNC and other devices such as a printer or landline modem.

A cheap terminal can be used to monitor activity when the computer is going to be busy for awhile. A connect alarm which beeps when someone connects to your station is a useful feature to alert you to switch over to the TNC. Connect alarms are available for some TNCs as a built-in function or as an add-on kit.

If the computer is shared with non-hams, be sure to consider the possibility of their accessing the TNC and getting on the air. It would be a good idea to disable the TNC's transmit OK command (in TAPR = XMITOK) when not using the TNC. This system allows you to access the TNC from a remote location which at first may seem to be an inconvenience, but it can be refreshing to get out of the shack and operate packet while sitting in the living room listening to the evening news.

Portable Station

Portable operation with packet radio is very useful. It is a lot of fun, and can be used while on trips or vacations. Emergency communications is an area of amateur radio which would benefit enormously by applying packet technology in the field. It is possible to fit an entire packet station with smart terminal, TNC, and radio in an attache case with room left over. (See Fig. 5-6.)

The basic requirements for a portable packet station are that it must be truly portable, consume as little power as possible, and possess full operating capabilities. A portable computer makes an ideal portable terminal. Units such as the TRS-80 Model 100 (see Fig. 5-7) feature an LCD display, full-size keyboard, built-in

Sample Stations 99

Fig. 5-6. A complete portable packet station in a briefcase consisting of a Model 100, a Pac-Comm TNC-200, an IC-02AT, and a gel cell battery.

Fig. 5-7. The TRS-80 Model 100 portable computer.

telecommunications software, on-board 300 baud modem, an RS-232 port, a printer port, and full programmability in BASIC or other languages with additional software. All of this capability fits in a case the size of a three-ring binder and weighs about the same. The entire computer is powered by 4 AA batteries for up to 20 hours. Supplemental power supplies are also available that allow for extended operation.

The Model 100 is very popular with packet operators. Its built-in telecommunications software features Xon/Xoff, download and upload, echo to printer port, and high baud rates. Many optional accessories are available, including a battery operated disk drive and printer, many software packages, and memory add-ons.

Other portable computers can be used with equal success. Also, a battery-operated hardcopy terminal is possible. These units often feature built-in acoustic modems for use with the telephone system. If it has an RS-232 port, it can be connected to the TNC and used as a terminal. However, these units are usually classified as dumb terminals.

The TNC should draw as little power as possible. Two hundred and fifty mA should be the maximum current drain considered for a battery operated portable station. Some manufacturers offer a version of the TNC with CMOS chips which draw much less power than their NMOS counterparts. They also generate less heat which is especially advantageous in a crowded portable arrangement. An on-board modem is very convenient as it reduces the number of components and the amount of wiring and interfacing needed.

For VHF operation, a handheld is the most obvious choice. They are compact, battery operated, and can generate up to 5 watts of power. Some require a special interface in order to allow the TNC to control the PTT. An amplifier can be added for additional range. However, the amplifier may significantly increase the T/R and R/T time of the radio as well as battery drain. Any antenna system commonly used for FM phone will probably work successfully on VHF packet. Of course, when operating at a fixed location for a while, you will probably want to put up an antenna with higher gain characteristics. I have had good results with a regular rubber duckie antenna running about 2 watts in dense urban areas where digipeaters are readily available.

Batteries are the most obvious power supply for a portable packet station. Twelve volt dc gel-cells, while rather heavy, offer 3 to 6 amp hour ratings and can be used to power most any station. (Be sure to carry a battery charger!) The same power system can be used to run the TNC, radio, and computer.

Radio Frequency Interference 101

HF portable operation is not very common in packet radio except at fixed sites. Power requirements, size, and antenna space are the major restrictions to portable HF operation. Most of the HF rigs that are used for packet operation are modern transceivers such as the Kenwood TS-430S.

Packet radio can be adapted to fit in with most any shack and operating situation. Setting up a packet station is not difficult. It just requires a little thought before running out and gathering equipment.

RS-232C

Since the RS-232 standard is utilized so much in packet radio, an appendix has been included which covers the various peculiarities involved in interfacing equipment. If you run into problems when connecting your TNC and terminal, check the appendix for possible solutions.

RADIO FREQUENCY INTERFERENCE

One problem which many amateur radio operators who work with computer equipment in the shack experience is *Radio Frequency Interference (RFI)*. Just as computer equipment can interfere with TVs and stereos, it can interfere with amateur radio equipment. RFI can be a maddening problem, and a logical approach is necessary to identify and correct it.

Even if you haven't experienced RFI, it is a good idea to take note of all the computer equipment in your shack. Of course, your desktop micro is the best place to start, but don't forget your TNC and any other devices which utilize digital circuits. You may experience RFI later on and immediately blame it on your micro, only to discover that the culprit was an easily overlooked piece of equipment, such as a digital clock.

RFI may be experienced on the HF and/or VHF bands. RFI is generated by computer equipment's internal signals. Computer signals are in the form of square waves which are composed of many sine waves. Because computer signals are switched so quickly, the sine waves cover a broad bandwidth. If the computer is not well-designed and lacks adequate shielding, the computer-generated RF can leak out of the computer system and invade other electronic equipment in the shack, or elsewhere in the house.

TNCs are usually designed to eliminate RF interference problems because of the sensitive on-board modem section. This is not to say, however, that a TNC should not be suspected of causing RFI. The FCC now regulates RFI levels for manufacturers of computer equipment. Some early micros, such as the TRS-80 Model I, were not well-shielded, so they generate a great deal of interference.

If you suspect you are experiencing RFI, the first thing to do is to identify where it is coming from. Turn off each piece of computer equipment in the shack, one at a time, and listen for the interference to stop. Before you turn off a computer with a keyboard, type on it, and see if you notice a change in the interference signal corresponding to your typing. If you find the trouble lies with the computer, power it up and make sure the interference returns. Try powering up one component at a time and isolate the offending component. If possible, substitute for the component to see if the fault lies with your particular unit.

If you find the interference is coming from the main CPU unit, disconnect all peripherals except for the TNC. Check for interference again. If there is still interference, disconnect the TNC. If the interference disappears, then it must be coming out the TNC cabling.

The first step for RFI reduction is to rearrange the station. Move the computer equipment to a separate electrical circuit. Check the computer equipment's case. Is it well shielded? Is it made of metal or plastic? Check the PC board. Is it grounded? Do the CPU and related components have a metal housing around them?

If the case has inadequate shielding, try coating the inside of the case with aluminum foil or tape, copper plating, or conductive spray. If you suspect the interference is leaking through the cables, try varying the length of the cable; it may be acting as an antenna. If the cable is not shielded, replace it with shielded cable or wrap it with foil. These solutions can also help protect the computer equipment from outside interference.

RFI can be a difficult problem to solve. Always start out with easier solutions and work your way up to the more complex. Check with other owners of similar equipment for possible solutions. There is a pamphlet available from the Government Printing Office that deals with RFI: "How to Identify and Resolve Radio-TV Interference Problems," Stock No. 004-000-00345-4. You can order it from the US Government Printing Office, Washington, DC 20402. Also, write the manufacturer of your computer equipment. They may have information or suggestions to help you.

ANTENNA SYSTEMS

Thus far, we have been concentrating on the most obvious parts of a packet station such as terminals, TNCs, radios, and interfaces. However, once the modulated RF has left the transmitter, there are several steps that you can take to maximize its range and quality. Your station's reception can also be improved.

In VHF operation, antenna height is always an advantage. A vertical omni-directional antenna can be used for monitoring, and a directional antenna can be switched in when operating. This will allow you to monitor from all directions, yet when operating, use no more power than necessary, have a directed signal, and let other stations use the frequency at the same time. A strong omni-directional station will be copied by many stations over a very wide area. Depending on your local operating characteristics, it may be to your advantage to increase your range beyond the station you are communicating with if you find you are being transmitted on top of by a strong station that cannot hear you. This will enable the carrier sense of the stronger station and allow you to have uninterrupted transmissions.

A directional antenna's gain characteristics and limited field of view can help when receiving a weak signal. Preamps can be useful when dealing with weak signals, but beware of possible distortion. A good deal of your VHF antenna needs will depend on local conditions. How far is it to the nearest station? Is there a strong digipeater which you pick up, yet can't reach?

For HF operation, no special antenna systems are usually needed. It is important to have a strong signal on HF to counter the interference found on most bands. On HF, it is not possible to be as selective in coverage and range as it is on VHF. A good, stable transceiver is your best plan of attack for HF operation.

Receive diversity is a concept that can be applied to both HF and VHF operation. In a receive diversity system, two antennas are located apart and connected to the same radio through a splitter. This can help improve reception, especially on HF with changing propagation conditions. It is always a good idea to maximize the performance of the receive section; regardless of the specific methods used. A good receive section can reduce the number of retries due to reception errors.

While amplifiers can increase your signal range and quality, they can also greatly increase the T/R and R/T times of your radio. Also beware of possible distortion, as with any component added in the signal's path to the modem. Using good-quality, low-loss coax is another way to reduce the loss of your station's antenna system and maximize your power output. Another possibility is to locate the amplifier at the antenna.

Chances are good that your regular antenna setup will work fine. Remember that vertical polarization is used on VHF. Use it as is, and adapt it to packet operation as you find necessary. For more detailed information, check a book on antenna system design. Concentrate on the other areas of your station that require more adaptation for packet operation first.

CONCLUSION

In this chapter, we have looked at the many variables that go into setting up a packet radio station. If the various choices are made logically with an idea of the complete system in mind, there will be few regrets. The cost of setting up a packet station varies widely depending on what equipment you already have and what kind of a station you want to have.

Take a look at what you have and what you need. Research the available equipment for items that meet your needs. Try to visit a local packet operator to get some more ideas and learn more about your local packet operating conditions.

Now that you know about setting up a packet radio station, the next chapter discusses some ideas about how to use it.

Chapter 6

Operating Amateur Packet Radio

Getting a packet station put together is only half the battle when first getting started in packet radio. Packet is a unique mode with its own operating characteristics. New users are usually somewhat apprehensive about operating packet radio because of the strange terminology and operating practices. Packet radio is simply a means of communicating information through a versatile radio channel very efficiently. As long as you keep this in mind and do not become overly-concerned about the new operating procedures, you will quickly pick them up while learning to enjoy the capabilities of packet.

What you learn here will save you from "on the air" trial and error. After a while, packet becomes second nature, and new users stop thinking in terms of terminals, ASCII, RS-232, modulation methods, encoding techniques, and radios. Packet becomes a system and they realize their station, with all its individual components, is just a small part of a large network.

This chapter starts off with a discussion of user interfaces. The user interface is the first obstacle you encounter once you get your station operating. After the introduction to user interfaces, initial parameters that usually have to be set before operation can begin are covered. Then the processes of initiating connections, conducting QSOs, and disconnecting are discussed. Bulletin board and other packet systems are covered next. The chapter closes with a list of some packet operating tips which may help you utilize packet radio more effectively.

THE USER INTERFACE

The user interface includes the command set, terminal specifications (such as screen size, baud rates, and control codes), and structure and design considerations (such as menus, modes, and prompts). Most TNCs have divided their user interfaces into three different areas known as modes. Each mode serves a different purpose. The command mode is used when entering commands and changing parameters. The conversation mode is used while transmitting and receiving information and features many editing functions for preparing your data. The transparent mode is also used while transmitting and receiving information, but it does not have any editing functions; the TNC is "transparent" to the user. The transparent mode can be used to send transmit and receive binary programs and other data which might trigger the converse mode's editing functions.

The user interface will be your window to the packet world, so you should spend some time learning about its features. The ideal user interface should be easy to use, learn, and understand. In addition, it should be flexible enough to allow the user to skip over areas that aren't of interest while simultaneously providing extensive capabilities in other areas.

Current packet user interfaces come very close to the ideal; much thought and experimentation have gone into their design. There are two main classifications of user interfaces that can be made. The first is for universal TNCs, and the second is for specific TNCs.

A universal TNC is a TNC designed for use with a wide variety of terminal configurations (i.e., they work with almost any terminal). Because of this, their user interfaces are rather generic in design. They do not exploit the features of any one type of terminal in an effort to ensure a wide degree of compatibility.

There is little or no menu utilization, and the displays are usually very simple. They feature full command sets and one-to-three different operating modes. Prompts usually consist of a single line, and displays are not formatted in any special manner.

On a specific TNC, the user interface is adapted for use with a specific terminal (or computer). It often takes advantage of any special features that the terminal may have, such as graphics capability and function keys. Because the screen size and other variables are known, menus and complex displays are often used. A full command set is offered, but because of the structured design of the interface, certain commands may only be accessed in certain sections. For example, transmitter parameters may only be accessed when in the radio interface menu.

The command set includes all commands and other parameters that can be utilized by the user. There are two types of commands: those that are executed immediately and then forgotten, and parameters that are altered and retained for future reference. For example, the connect command is executed and must be reentered when you want to connect to another station. But the retry command is a parameter which is set by the user and is recalled each time a frame is sent.

The TAPR command set is the most prevalent in packet radio today, and has been widely copied in the majority of TNCs. It is the standard in much the same way as the Hayes command set has become the standard in the landline modem world. (Some other command sets have been developed for use with particular TNCs, such as the WA8DED code for the TNC-1 and compatibles, and the GLB command set.) The major benefits of the TAPR command set are that it is easy to learn and use. It features a full English syntax along with optional abbreviations. (See Fig. 6-1.) For example, the command to connect to KR3T is "Connect KR3T," or it can be abbreviated to just "C KR3T." The TAPR command set has three modes: the command mode, the converse mode, and the transparent mode.

In the command mode, most commands are followed by an argument. For example, in the "Connect KR3T" example, "Connect" is the command and "KR3T" is the argument. There are two types of arguments, toggle and response. A toggle argument switches between two different conditions. The toggle argument in the TAPR command set consists of "On" or "Off." The second type of argument, response, consists of a string of characters as in the "Connect KR3T" example in which "Connect" is the command and "KR3T" is the response string or argument. After typing a command and argument, a key (usually RETURN or ENTER) must be pressed. This lets the TNC know that the command is finished and ready to be processed. If an error has been made when entering the command, the TNC displays an error indicator and redisplays the command prompt.

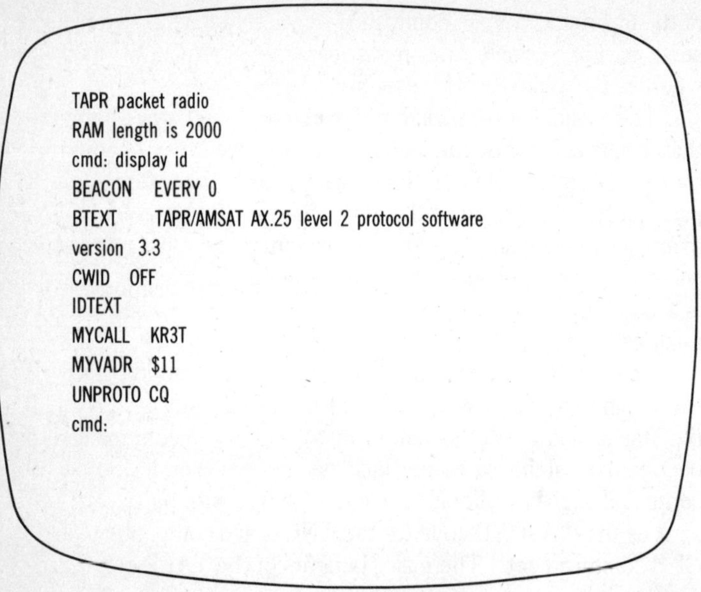

```
TAPR packet radio
RAM length is 2000
cmd: display id
BEACON    EVERY 0
BTEXT     TAPR/AMSAT AX.25 level 2 protocol software
version 3.3
CWID   OFF
IDTEXT
MYCALL  KR3T
MYVADR  $11
UNPROTO CQ
cmd:
```

Fig. 6-1. The TAPR user interface. The **DISPLAY** command is used to display parameters. In this case, only the station's ID parameters were selected.

INITIAL PARAMETERS

There are several commands that must be set to configure the TNC so that it will work properly with the rest of the station. These commands can be divided into three categories: terminal parameters, radio parameters, and operational parameters. Terminal parameters are those commands which must be set to the proper values necessary for communications with the terminal. Radio parameters are those commands that must be set to the proper values necessary for interfacing with the radio. Operational parameters are those commands which must be set to the proper values for specific operating conditions.

The following section lists the commands that must usually be configured before going on the air. Not all TNCs have all the commands listed, and many have more. However, this section gives you an idea of what parameters need to be set, and specific references are made to the TAPR and GLB command sets. In most TNCs, once a parameter is set it is permanently retained so that it is not necessary to reconfigure the station each time it is turned off. Some even retain more than one set of parameters so it is easier to configure the TNC for different operating conditions such as HF, VHF, and satellite.

Terminal Parameters

Some of the terminal parameters deal with configuring the RS-232 serial link between the terminal and the TNC. Refer to Chapter 1 for more information on the specifics of the ASCII and Baudot codes and asynchronous serial communications, and to Appendix B for more information on the RS-232-C interface.

The first terminal parameter is the baud rate, the rate of data transfer between the TNC and terminal. In this case, the baud rate is equal to the number of *bits per second (BPS)*. Do not confuse the terminal baud rate with that of the radio link. The terminal baud rate on most TNCs is capable of running at up to 19,200 BPS. The terminal baud rate should be set to that of your terminal. The default value is usually 300 baud on most TNCs.

Some TNCs have an autobaud routine which monitors the output of the terminal and automatically adjusts to match it. Once the terminal baud rate is set, the TNC must be reset before it will communicate at the new rate. The TAPR command for the terminal baud rate is ABAUD.

The next terminal parameter is the word length. This is the number of bits per character. It is set to either 7 or 8 for ASCII and 5 for Baudot. Most ASCII terminals only recognize 7 bits unless told otherwise. Eight bits allow for special characters not included in the regular 128 ASCII character set. The default is usually 7 bits per word. The TAPR command for the terminal word length is AWLEN.

Another terminal parameter is the number of stop bits to follow each character, either 1 or 2 bits. With an 8 bit word length, only 1 stop bit is usually allowed. The default is usually 1. The TAPR command for the number of stop bits is ABIT.

The final serial interface terminal parameter is the type of parity desired. Parity is a form of error detection which counts the number of bits in a character and sets an additional bit, called the parity bit. There are two types of parity; even and odd. In even parity, if the number of 1 bits in a character is even, the parity bit will be set to 0; if odd, it will be set to 1. In odd parity, the opposite is true—if the number of 1 bits is odd, the parity bit will be set to 0; if even, the parity bit will be set to 1.

It is also possible to ignore parity by setting the parity parameter to "none" or "off." If no parity is requested, the TNC should not even include a parity bit in the transmission. However, in some cases, the parity bit may be set continuously at either 0 or 1 and simply not checked. The TAPR command for setting the type of parity desired is PARITY.

If the default TNC parameters do not match those required by your terminal, simply type the appropriate command and the desired argument in the command mode and it will be changed. In most cases, it is best to start out with a 1200 baud terminal baud rate, a 7 bit word, a 1 bit stop bit, and even parity. Be sure your terminal parameters are set to the same values as the TNC. If your terminal doesn't seem to be working, try experimenting with different parameters.

Machine-specific TNCs and software TNCs usually do not use an asynchronous serial port for terminal to TNC communications, so they will probably not have these terminal commands.

The following TNC parameters can be classified as terminal characteristics. These commands configure the output from the TNC to your terminal.

The first of these parameters is the screen width. This is the number of characters the TNC will send before adding a linefeed character. Set this to the character width of your display device. The TAPR command for setting the screen width is SCREENL. The default is usually 80 characters per line. Even if your display is not 80 characters, you might want to leave the screen length set to 80 so that printouts of received text are properly formatted, since most people use 80-column display devices.

The next parameter controls the number of nulls sent after each linefeed or carriage return. This allows time for the terminal to get ready for the next line. A null is a blank character which causes the terminal to do nothing. Most CRT terminals do not require any nulls. However, in the case of hardcopy terminals, nulls may be required after each carriage return to allow time for the mechanism to return the printhead to the beginning of the next line. A slow CRT terminal may require nulls after a linefeed to allow time for the terminal to update the video display.

Initial Parameters 111

In the TAPR command set, the command to activate the sending of nulls is NULLS (ON/OFF); NULLS is a toggle command. The Commands NUCR and NULF are used to set the number of nulls to be sent after a carriage return character and linefeed character respectively.

With the echo parameter, it is possible to configure the TNC to echo characters received from the terminal back to the terminal for display. Note that this is not the same as echoing characters to a printer from a terminal. If the terminal is set for full duplex operation and the TNC is set to echo, all characters typed are sent back to the terminal for display. However, if the TNC is set to echo and the terminal is set for half duplex, both the terminal and TNC send the character to the display and double characters will result. Either enable the echo feature and set the terminal to full duplex or disable the echo feature and set the terminal for half duplex. It is usually better to set the TNC to echo so you can be sure that the characters are being received properly. The TAPR command to set the echo is ECHO.

Another terminal parameter is the auto linefeed. If this is set, the TNC adds a linefeed character after each carriage return sent to the terminal. If the terminal also adds a linefeed after a carriage return, double spacing results. The TAPR command to add a linefeed is AUTOLF.

If your terminal does not support lowercase characters or if you wish to use case to differentiate between sent and received text on the display, the TNC can be configured to send only uppercase characters to the terminal. All lowercase characters received by the TNC are converted to uppercase before being sent to the terminal. The TAPR command is LCOK.

The remainder of the terminal parameters deal with control codes. These are special characters recognized and acted on by the TNC. Most are for editing in the converse mode. The default values usually work fine with most terminals.

The first control code is the backspace character. This is one of the most frequently used characters, as it is used to remove typing mistakes. It can be used in both the command and converse mode. There are actually two keys which may be used for correcting typing mistakes, backspace key and the delete key. The backspace key is represented on most terminals as the left arrow or BKSP. The delete key is represented (on the keyboards that have one) by DEL or DELETE. The TAPR command to choose between these two keys is DELETE. DELETE is a toggle command and ON selects the delete key; OFF selects the backspace key.

A hardcopy terminal obviously can't backspace over characters that have already been printed, so the TAPR command set allows for corrections on hardcopy terminals using a different method. The TNC can be configured to send a "/" to the terminal for each character deleted. On a CRT terminal, the TNC backs the cursor one space, displays a space, then backs the cursor again. To select between the two methods of updating the display the TAPR command, BKONDEL is used. If BKONDEL is ON, the CRT updating method will be used. For a hardcopy terminal with a backspace key, DELETE should be OFF and BKONDEL OFF.

Other heavily used control codes are the Xon/Xoff flow control codes, used to temporarily interrupt the data flow using Xon/Xoff software flow control. The control characters to be used can be set. The values should be set to those of the terminal. The default is Xon/Xoff enabled, the Stop character set to CTRL-S, and the Start character set to CTRL-Q. The TAPR command to toggle between Xon/Xoff enabled and disabled is XFLOW. The control characters are set with the START and STOP commands. The GLB command set does not include Xon/Xoff commands.

The command to set the control character to exit the converse or chat mode and return to the command mode is COMMAND for the TAPR command set. The default for COMMAND is CTRL-C.

The following terminal commands are used in the converse and chat modes to edit data being typed in for transmission.

The TAPR CANPAC command is used to delete the data entered since the last packet sent. The default value for CANPAC is CTRL-Y (HEX 19).

The TAPR CANLINE command is used to delete the data inputted since the last carriage return (CR) was typed. The default value is CTRL-X (HEX 18).

The TAPR REDISPLA command is used to set the control code used to redisplay the current line being inputted. The default value is CTRL-R (HEX 12).

The TAPR SENDPAC command is used to set the character used to send a frame. The information entered since the last SENDPAC is put into a frame and transmitted. The default is a carriage return <CR> (Hex 0D). Whenever a <CR> is pressed, a frame is sent.

That completes the terminal parameters and commands. Again, these parameters may only apply to universal hardware TNCs.

Radio Parameters

The next group of TNC parameters deal with the radio. They are designed to tell the TNC about the timing characteristics of the radio and network to which it is interfaced.

The most important radio parameter is the transmitter delay. This parameter tells the TNC how long to wait after keying the transmitter before sending the frame. This time is required by the radio to key up properly. In the TAPR command set, TXDELAY is the command used to set the delay. A number within a certain range is entered. That number is then multiplied by a time value to give the total delay.

The TAPR command set allows for an additional delay period for use with slow audio repeaters or receivers with a slow squelch release. This delay is the time to wait in addition to TXDELAY before sending data. The command is AXDELAY and the argument follows the same guidelines as TXDELAY.

There is one more TAPR command used in conjunction with AXDELAY which is used to negate the AXDELAY. This is necessary when using an audio repeater with a long squelch tail (meaning the repeater remains keyed up for a time period after repeating). Since the repeater is already keyed up, there is no need to wait AXDELAY before transmitting. If the TNC has heard a packet within a time period, it will not add AXDELAY to the TXDELAY. This time period is set by the AXHANG command. It uses the same arguments as AXDELAY.

This concludes the radio parameters included on most TNCs. Machine-specific TNCs and software TNCs also have radio parameters, as they do not contain integrated radios.

Operational Parameters

The last section of commands are the operational parameters. These commands are used to set the TNC to match your operating preferences and local operating conditions.

The first command is used to select the protocol you wish to utilize when operating. On older TNCs, the selection is between AX.25 and VADCG (V-1). On current TNCs, the selection is between AX.25 Version 1 and AX.25 Version 2. The TAPR (TNC-1 version) command to select between AX.25 and V-1 is AX25. AX25 uses a toggle argument with ON selecting AX.25 and OFF selecting V-1. The TAPR (TNC-2 version) command to select between AX.25 V1 and AX.25 V2 is AX25L2V2. This is a toggle command also with ON selecting Version 2.

The next command is used to select which of the communications modes is entered upon establishing a connection. The choice is between converse and transparent; converse is almost always chosen. The TAPR command to select which mode is to be used is CONMODE. The two arguments are CONVERS and TRANS; CONVERS is the default value.

The TAPR command CONOK is used to indicate if your station is available for connection. CONOK is a toggle command with ON selecting connection availability. If CONOK is set at OFF, the originating station will receive a busy message. The CONOK default is ON.

The next command is similar to CONOK but is for use in allowing other stations to use your station as a digipeater. If your station is available for use as a digipeater, the command should be enabled. In the TAPR command set, there are two digipeater commands, one for AX.25 and one for VADCG (V-1). The TAPR (TNC-1 version) commands are DIGIPEAT (ON/OFF) for AX.25 and VDIGIPEA (ON/OFF) for V-1. Only one TNC in a particular area should be activated as a V-1 digipeater. Check with other local operators before you activate the VDIGIPEA command.

The TAPR command FLOW controls the flow of received data from the TNC to the terminal. The received data can be sent to the terminal immediately upon reception, or they can be held if the user is in the process of typing a line until the carriage return key is pressed. This command works in both the command and converse modes. This command keeps the screen clearer but takes away the advantage of rapid interaction between users. The TAPR command is FLOW (ON/OFF) with the default ON limiting the flow of received data as described above.

The TAPR (TNC-1 version) command HBAUD is used to set the link baud rate. This is the baud rate of information sent to and received from the modem. Remember that Bell 202 is 1200 baud and Bell 103 is 300 baud. The TAPR command HBAUD is set according to a table of values. The default is 1200 Baud.

The next four commands are used to enter your station's address. In AX.25, the address of your station consists of your callsign plus a substation ID number between 0 and 15. In VADCG

(V-1) the address consists of a number between 0 and 31. Some TNCs contain the user addresses in ROM, so it may not be necessary to enter yours. The TAPR command to enter your AX.25 address is MYCALL. The TAPR (TNC-1 version) command to enter your VADCG address is MYVADR. Newer TNCs may contain a second AX.25 address when using the TNC as a digipeater. This second address may be set to an area code, airport identifier, or anything else (it must be six characters or less).

The TAPR command XMITOK can be used to disable the TNC from transmitting. It is useful if the station is left on and unattended for long periods of time to prevent someone from accidentally keying the transmitter. XMITOK is a toggle command and the default is ON meaning it is OK to transmit.

The remaining TNC commands are used to configure the timing and other specifics of the TNC, and it will probably not be necessary to change them from their default values until you have a solid working knowledge of the networking conditions in your area.

The DWAIT command is used to reduce congestion on the channel by allowing digipeated packets first chance at the channel. Thus the digipeated packets are less likely to be collided with and then will not have to be sent all over again from the originating station. All stations except digipeaters have to wait the time period set with DWAIT before transmitting. A digipeater that is ready to transmit at the same time as a user will capture the channel first. The DWAIT command is set in small time intervals. The DWAIT should be set to a little over the TXDELAY of most transmitters (i.e., DWAIT = 100 to 200 ms).

The RETRY command sets the number of times a frame is to be retransmitted before unilaterally disconnecting. The TNC sends a frame a total of this value plus 1 since the first transmission was not a retransmission. For example, if you want only 15 attempts at sending a frame before disconnecting, the value should be 14. The TAPR command is RETRY and the default is 10. A 0 means unlimited retries.

The next command, FRACK, sets the length of the wait after sending a frame before it is retransmitted if no acknowledgement is received. If the path of the connection includes digipeaters, it takes longer for the frame to reach the destination station and for the acknowledgement to return than if the two stations were linked directly. Because of this, the total delay is adjusted for each digipeater. The formula for determining the total delay is FRACK*((2*# of digis)+1). Thus, if there are no digipeaters in the path, the total delay equals the FRACK value. In the TAPR command set, the FRACK value is set in increments of 1 second.

The MAXFRAME command sets the maximum number of frames that can be transmitted at a single time. There is an upper limit of 7 under AX.25. If some of the frames in a transmission are acknowledged, they are dropped and that many more added to the next transmission if they are available.

The last command sets the maximum number of bytes (or characters) of user data that can be included in a frame. When the keyboard input reaches this amount, the user data is automatically put into a frame and sent. The TAPR command PACLEN can range from 1 to 256 bytes. The default is 128 bytes. Having more than 128 bytes of information is an extension of the AX.25 and V-1 protocols, so values greater than 128 may not work with all TNCs.

Together with the maximum number of frames command, it is possible to control the amount of data in each transmission. The maximum amount of data is 7 frames with 256 bytes of user data in each. Short transmissions are more likely to get through without error in conditions of interference or congestion as found on HF. Long transmissions are more efficient on reliable, open channels.

That concludes the initial parameters needed to configure your TNC. By now you know what packet radio is, how it works, how to set up a station, and how to configure the TNC. The next step is to actually get on the air and operate amateur packet radio.

MONITORING

One feature available on all TNCs that is very useful when first getting on the air as well as for regular operating is monitoring. Monitoring allows you to monitor activity on a channel by disassembling all frames received by your station that meet certain conditions that you specify. In other words, you can select the activity you want to monitor. The general TAPR command for monitoring is MONITOR. It is usually a toggle command with ON selecting monitoring. However, other TNCs may implement a numeric response code for an argument which selects the type of monitoring desired. It is usually possible to selectively monitor only frames containing user information, control frames, digipeated frames, or frames with specified addresses.

The first connection you should attempt is to yourself through a digipeater. This allows you to check your station and get used to the process of initiating a connect, transferring information, and disconnecting. But before you can connect to yourself, you need to locate a digipeater that you are able to use. Digipeaters are usually easily located by monitoring. Look for a frame sent as unnumbered information to a general address such as "BEACON" which states that the station is available for use as a digipeater. Some digipeaters may not advertise their availability, so monitor the activity of other stations in your area.

Beacons

Now is a good time to introduce the subject of beacons. As beacon is a frame sent as unnumbered information. An unnumbered information frame is sent (broadcasted) on the channel for all users to receive. It can be sent regardless of whether the TNC is connected to another station. Because there is no destination address, a user-selected dummy address is placed in the address field of the frame. The dummy address is usually BEACON, although CQ is sometimes used.

In the TAPR command set, the text to be sent in a beacon is entered using the BTEXT command. Beacons can be sent periodically or only after activity is heard on the channel. The command to activate the sending of a beacon is BEACON. The argument for a beacon to be sent repeatedly every specified time period is EVERY # (# = number of time intervals). The argument for a beacon to be sent after a specified pause in activity is AFTER#. The TAPR command to set the dummy address used by the beacon is UNPROTO and can include up to eight digipeaters.

Unnumbered frames can also be broadcasted manually by entering the converse mode with the CONVERSE command and inputting the text. This can be useful when you want to transmit information to a number of users at a single time, such as in a roundtable discussion. However, there is no guarantee that the frames will be received correctly by all stations as there is no error-checking. When the send packet character is pressed, the frame will be broadcasted using UNPROTO as the destination address.

Just because all TNCs have the capability to send beacons does not mean that they should be used by all users. Beacons should be reserved for information that is of importance to a large number of operators. They should always be kept as short as possible, and the time period between transmissions should be kept as large as possible.

Connecting to Your Terminal

Once you have located a digipeater by monitoring or other means, you should try to connect to yourself through it. All commands are generic TAPR because of its popularity. The command to connect to another station through a digipeater is CONNECT call VIA digi. Up to eight digipeaters can be used and are separated by commas. For example, suppose I want to connect to myself using KR3T-1 as a digipeater. I type CONNECT KR3T VIA KR3T-1 at the command prompt; the abbreviated form is C KR3T V KR3T-1.

My TNC transmits a connect request frame which is received by KR3T-1 if it is in range and operating. KR3T-1 sees that its address is in the digipeater field and resends the connect request. My station, KR3T, receives the connect request and sends an acknowledgement to KR3T which is digipeated by KR3T-1. Once I receive the acknowledgement, I am connected to myself and automatically put in the CONMODE, usually converse.

Anything I send is sent back to me by KR3T-1 (See Fig. 6-2.) Typing a CTRL-C puts me back in the command mode where I type DISCONNECT. After the disconnect request is sent via the same path as the connect request and acknowledged, I am disconnected from myself. More information on connecting through digipeaters is given later in the chapter.

Connecting to Another Station

Now that you are sure the station is operating properly because you were able to connect to yourself, you can connect to another packet station and conduct a packet QSO. The first example shows how to connect to another station without the use of a digipeater, and the second demonstrates the use of multiple digipeaters.

Make sure your station is properly set up (correct frequency, your call entered, CONOK ON, and XMITOK ON). Look for the CMD: prompt. If it is not visible, press the <CR> key, and if that doesn't work, type the COMMAND character (usually CTRL-C) or the Xon START character (usually CTRL-Q). As a last resort, turn the TNC off and on again. Pick

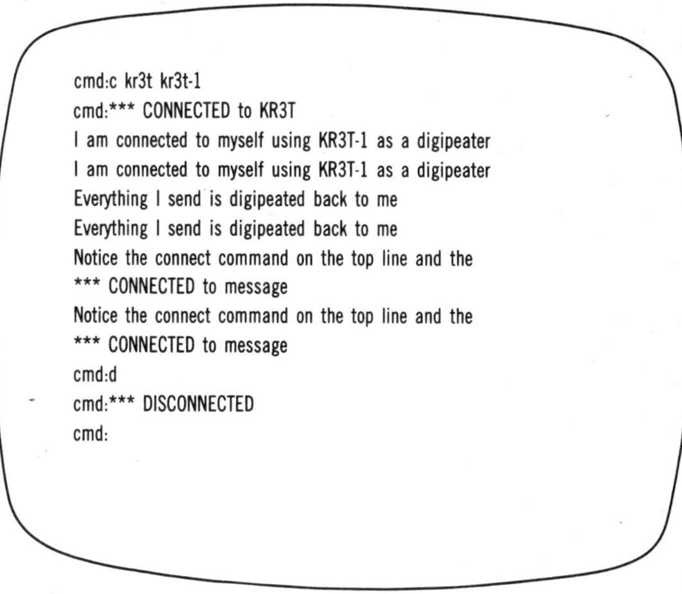

```
cmd:c kr3t kr3t-1
cmd:*** CONNECTED to KR3T
I am connected to myself using KR3T-1 as a digipeater
I am connected to myself using KR3T-1 as a digipeater
Everything I send is digipeated back to me
Everything I send is digipeated back to me
Notice the connect command on the top line and the
*** CONNECTED to message
Notice the connect command on the top line and the
*** CONNECTED to message
cmd:d
cmd:*** DISCONNECTED
cmd:
```

Fig. 6-2. Example of a self-connection using a digipeater.

a station to connect to that is within range of your station, possibly one that you have recently monitored.

The following example assumes you are trying to connect to my station, KR3T-1.

Type C KR3T-1 at the CMD: prompt, which means send a connect request frame to KR3T-1. Then press the <CR> or RETURN key. The transmitter should key up briefly as the connect request is transmitted. There are three possible outcomes to a connect request.

The first is that you will *retry out*. This means that your TNC retransmitted the connect request RETRY number of times without receiving a response. This indicates that the station you are attempting to connect is out of range or not operating properly. The retry message will look like this: "***Retry count exceeded." (See Fig. 6-3.) You are then unilaterally disconnected and returned to the CMD: prompt.

The second possible outcome is that you receive a busy message from the station you tried to connect to. This looks like "***KR3T-1 busy." (See Fig. 6-4.) This indicates that I am already connected to another station or that I have my CONOK command turned OFF.

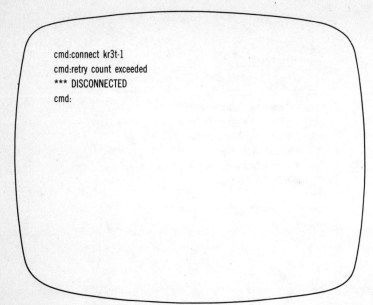

Fig. 6-3. The "retry out" message.

Fig. 6-4. The "busy" message.

The third possible response is that I acknowledge your connect request, and we will be connected. We are each sent a "***Connected to Call" message (you see my call and I see your call) and put in the converse mode by our TNCs, assuming that is what we have each set our CONMODE to. (See Figs. 6-5 and 6-6.) Now everything that we type on our terminals is put into frames and sent to the other station. The TNC keeps sending the frames until it receives an ACK from the other station. If no ACK is received and the frames has been retransmitted RETRY number of times, the sending station retries out.

The frames received by each TNC are disassembled and the information portion sent to the terminal. (See Figs. 6-7, 6-8, and 6-9.) It is possible to return to the command mode during a QSO and alter or check parameters and initiate commands by typing the COMMAND character (usually a CTRL-C). The CMD: prompt is then displayed. When you are through and want to return to the QSO, type CONVERSE at the CMD: prompt, and the TNC switches to the converse mode again.

Fig. 6-5. The connected message at KR3T.

122 Operating Amateur Packet Radio

Fig. 6-6. The connected message at KR3T-1.

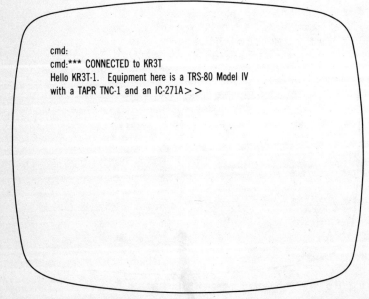

Fig. 6-7. Received frames from KR3T.

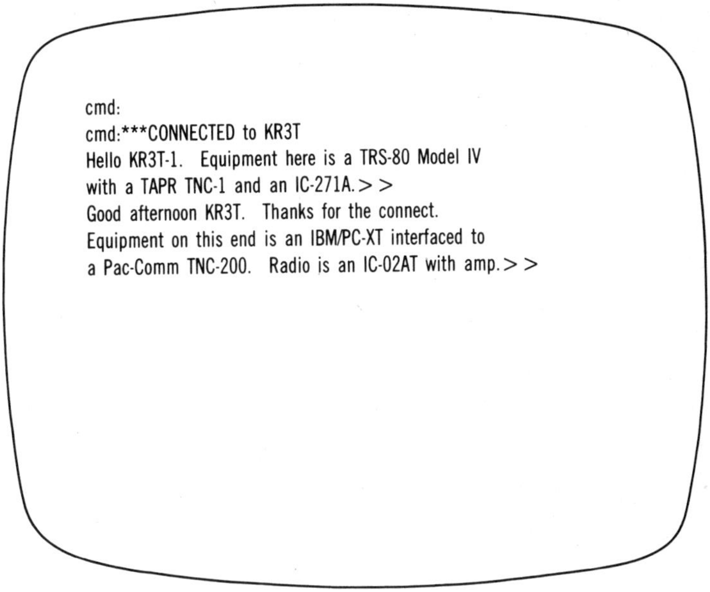

```
cmd:
cmd:***CONNECTED to KR3T
Hello KR3T-1.  Equipment here is a TRS-80 Model IV
with a TAPR TNC-1 and an IC-271A.>>
Good afternoon KR3T.  Thanks for the connect.
Equipment on this end is an IBM/PC-XT interfaced to
a Pac-Comm TNC-200.  Radio is an IC-02AT with amp.>>
```

Fig. 6-8. Additional text to KR3T.

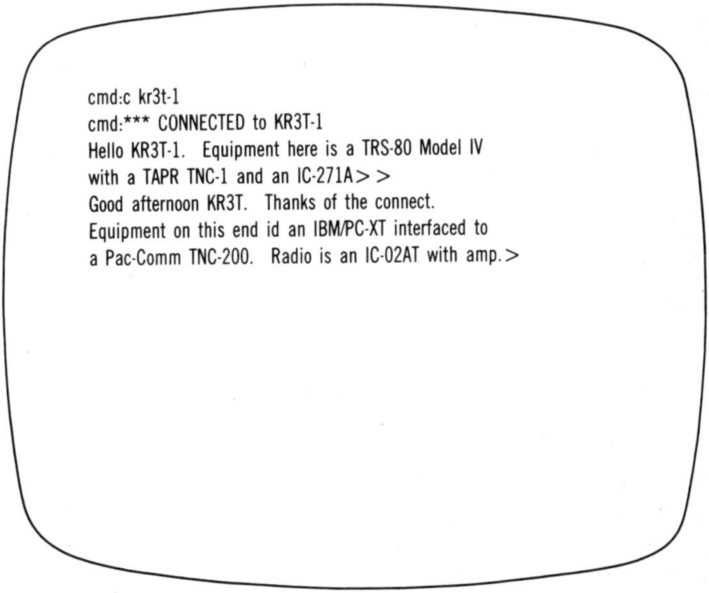

```
cmd:c kr3t-1
cmd:*** CONNECTED to KR3T-1
Hello KR3T-1.  Equipment here is a TRS-80 Model IV
with a TAPR TNC-1 and an IC-271A>>
Good afternoon KR3T.  Thanks of the connect.
Equipment on this end id an IBM/PC-XT interfaced to
a Pac-Comm TNC-200.  Radio is an IC-02AT with amp.>
```

Fig. 6-9. Text sent to and received from KR3T-1.

When you are through with the QSO (see Figs. 6-10 and 6-11), either station may initiate a disconnect by returning to the command mode and typing DISCONNECT (abbreviated DISC or just D) at the CMD: prompt. A disconnect request is sent to the other station and if acknowledged, the stations are each disconnected. If the other TNC

```
cmd:c kr3t-1
cmd:*** CONNECTED to KR3T-1
Hello KR3T-1. Equipment here is a TRS-80 Model IV
with a TAPR TNC-1 and an IC-271A.>>
Good afternoon KR3T. Thanks for the connect.
Equipment on this end is an IBM/PC-XT interfaced to
a Pac-Comm TNC-200. Radio is an IC-02AT with amp.>
FB. Well, TNX for the quick chat. I have to run.
U Disc. 73>>
```

Fig. 6-10. Additional text to KR3T-1.

does not ACK the disconnect request, it is retransmitted up to RETRY number of times. You are then unilaterally disconnected. It is also possible for you to initiate a unilateral disconnect by typing the DISCONNECT command a second time at the CMD: prompt; however, unilateral disconnections are not good operating practice.

When you are disconnected from another station the TNC sends a "*** Disconnected" message to your terminal. (See Figs. 6-12 and 6-13.) You are now able to connect to another station. Some TNCs allow a user to be connected to more than one station at a time. Each connection is assigned a logical channel (or stream), and the user can switch between QSOs by selecting the proper channel. The ability to connect to more than one station simultaneously is known as *multiconnect capability*.

Monitoring 125

```
cmd:
cmd:***CONNECTED to KR3T
Hello KR3T-1.  Equipment here is a TRS-80 Model IV
with a TAPR TNC-1 and an IC-271A.>>
Good afternoon KR3T.  Thanks for the connect
Equipment on this end is an IBM/PC-XT interfaced to
a Pac-Comm TNC-200.  Radio is an IC-02AT with amp.>
FB.  Well, TNX for the quick chat.  I have to run.
U Disc.  73>>
```

Fig. 6-11. Additional received text from KR3T.

```
cmd:
cmd:*** CONNECTED to KR3T
Hello KR3T-1.  Equipment here is a TRS-80 Model IV
with a TAPR TNC-1 and an IC-271A.>>
Good afternoon KR3T.  Thanks for the connect.
Equipment on this end is an IBM/PC-XT interfaced to
a Pac-Comm TNC-200.  Radio is an IC-02AT with amp.>
FB.  Well, TNX for the quick chat.  I have to run.
U Disc.  73>>
Roger.  Will try to connect with you later.  73
cmd:d
cmd:*** DISCONNECTED
cmd:
```

Fig. 6-12. Disconnect initiated by KR3T-1 and acknowledged by KR3T.

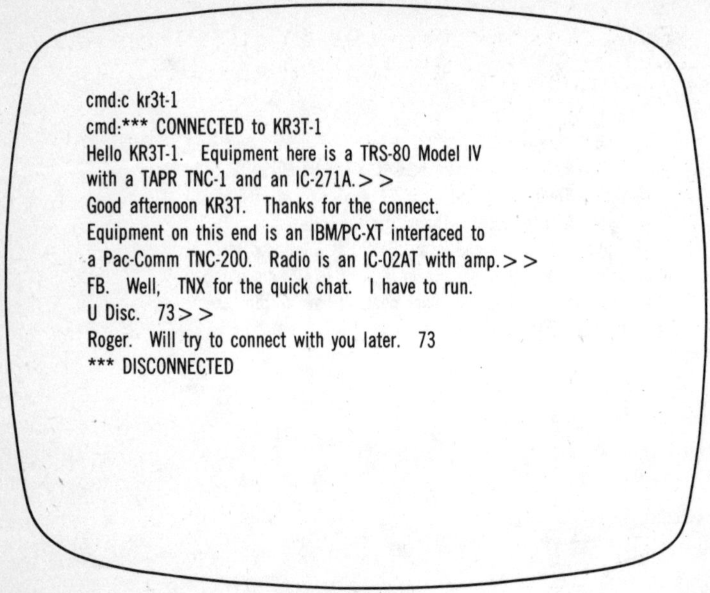

```
cmd:c kr3t-1
cmd:*** CONNECTED to KR3T-1
Hello KR3T-1.  Equipment here is a TRS-80 Model IV
with a TAPR TNC-1 and an IC-271A.>>
Good afternoon KR3T.  Thanks for the connect.
Equipment on this end is an IBM/PC-XT interfaced to
a Pac-Comm TNC-200.  Radio is an IC-02AT with amp.>>
FB.  Well,  TNX for the quick chat.  I have to run.
U Disc.  73>>
Roger.  Will try to connect with you later.  73
*** DISCONNECTED
```

Fig. 6-13. KR3T is also disconnected.

Connecting through Digipeaters

If the station you wish to connect to is out of range of your station, digipeaters can be used to bridge the gap between the two stations. The TAPR extension of AX.25 lets up to 8 digipeaters be used. The addresses of all digipeaters to be used must be known beforehand. To initiate a connect through digipeaters type C Call V Call, Call, . . ., Call (i.e., C KR3T V AA3F,KR3T-1). The connect request will be received by AA3F, digipeated to KR3T-1, and then digipeated to KR3T. Packets from KR3T will be sent to KR3T-1, to AA3F, and finally your station. The remainder of the QSO is identical to that of a direct connection.

In order to utilize digipeaters effectively it is necessary to learn a little about how they work. Most TNCs are capable of performing digipeater functions. A digipeater is a simplex packet relay station which receives all frames and checks to see if its address is contained in the digipeater address field. If the digipeater finds its address and it is its turn to relay the frame, the digipeater updates the digipeater address field to the next digipeater, if any, and then immediately retransmits the frame if the channel is clear.

Digipeaters do not retain copies of digipeated frames in memory or look for an ACK from the next station. The ACK for the digipeated frame must come from the destination station and be digipeated back to the originating station. Thus, if a frame is lost while being digipeated, the originating station must retransmit the frame from the beginning. This type of acknowledgement is known as an *end to end ACK*.

Digipeaters have many drawbacks. They are useful and beneficial to packet radio as long as they are used properly and responsibly. One user misusing digipeaters can cause heavy congestion on a channel. Each digipeater used increases the amount of channel time a frame consumes and the range of the transmission. The more times a frame is digipeated, the higher the chances that it will be lost along the way due to a collision or interference. The probability of getting a frame to its destination error-free on the first attempt is reduced geometrically as the number of digipeaters is increased.

Only use as many digipeaters as are absolutely necessary to maintain the link to the other station. Most active areas have several strong digipeaters that will allow you to reach most other local users in one or two hops. However, do not use more than two or three digipeaters on a moderately congested channel, and try to avoid digipeaters at all if the channel is very busy.

There are special digipeaters which are capable of receiving on one frequency and transmitting on another. These digipeaters are known as *dual port digipeaters*; the ports are on different frequencies. The dual port digipeater may be used as a regular digipeater by receiving and transmitting frames on a single port. The advantage of a dual port digipeater lies in its ability to automatically route frames that are received on one port or the other.

When a frame is received for digipeating, the next callsign in the address field is examined. If the callsign is recognized by the dual port digipeater as being on the other frequency, the frame is routed out on the other port. Alternately, if the dual port digipeater does not recognize the callsign, it examines the callsign's SSID. Based on a prearranged SSID-based system, the frame is routed out the appropriate port. This system allows traffic to be switched off the users' channel and digipeated on another, less-used frequency.

The maximum of eight digipeaters was added to allow remote areas to communicate over long distances. However, due to the rapid growth of packet radio it is almost impossible to utilize anywhere near eight digipeaters anymore in a moderately active area. Digipeaters are an admitted kludge to allow for rudimentary networking until the development of true level 3 network nodes. The use of digipeaters for long range communications will drop off significantly once these network nodes are implemented. See Chapter 4 for more information on the protocol specifics of digipeating.

More operating tips are given at the end of this chapter. Now we will turn our attention to one of the largest areas of amateur packet radio activity: bulletin board operation.

BBS OPERATING

Bulletin boards are known by a variety of nomenclatures including BBS for Bulletin Board Station, PBBS for Packet BBS or Public BBS, CBBS for Computerized BBS, and Mailbox. Bulletin boards open a new realm in amateur packet radio by providing nonreal time operations. It is possible for two users to access a BBS at different times and still transfer information.

All bulletin boards provide some form of message storage. A user may connect to the board and send a message to be stored in the board's memory or storage. Another user may then connect to the BBS later and read the message. Once a message is stored on a BBS it can usually be read an unlimited number of times. Thus, bulletin boards are an excellent method for sending messages to a large number of users.

Packet bulletin boards have existed almost as long as packet radio itself. And like amateur packet radio, bulletin boards have gained many new features and applications over the years. The current packet bulletin boards offer a wide variety of services for their users that greatly extend the usefulness of packet radio.

Just as the TAPR TNCs have become the *de facto* standard for design and user interfaces, one bulletin board system has emerged as the standard in amateur packet radio. This BBS system was developed by Hank Oredson W0RLI and is in use worldwide.

The first W0RLI BBS went on the air in February of 1984 in Massachusetts and has been spreading ever since. Hank's first BBS was modeled after an electronic message system that he had developed while working for Sperry Corporation. The system has been heavily modified over the years by Hank in response to comments and suggestions by users.

The original W0RLI BBS ran on an IMSAI system but was ported over to a Xerox 820 CP/M machine in March of 1984. The basic W0RLI configuration consists of a Xerox 820 computer board with keyboard, a monitor, one or two disk drives, and one or two TNCs with radios. The W0RLI system has been under constant development since it was first brought out in 1984. Starting at Version 2.0, it has progressed into a final form at Version 11.6. Although Hank expects to continue making minor revisions, he is finished with all hardcore modifications at this time.

The W0RLI system has been ported to many other computer systems besides the Xerox 820 with a high degree of compatibility. These W0RLI clones include an IBM-PC version written in Pascal by WA7MBL, a Commodore 64 version by WB4APR (not completely compatible), an OS-9 version by VE3FXI, a DEC Rainbow version written in Pascal by AK1A, and a version for the DEC PDP-11 by KA1T. Versions of the W0RLI system have also been written for the TRS-80 Model 100 and Kaypro computers.

The following section describes the W0RLI BBS in detail and gives examples of commands and operating procedures.

The W0RLI BBS system features a message system whereby messages can be stored on the BBS for retrieval by other users or forwarded to other BBSs. There are files available for downloading such as general information (for example, user lists and newsletters) and computer programs. There is a provision available on many W0RLI BBSs for linking to other frequencies through the BBS; these are known as *gateways*. A frequency activity monitor and message beacon which lists those stations who have messages waiting for them on the BBS are also included.

To use a W0RLI BBS, just connect to the station as if it is a regular user station. Digipeaters are fine although some BBS owners (known as SYSOPs for SYStem OPerator) have imposed a limit on the number of digipeaters that may be used. The SYSOP's callsign is usually the address of the BBS.

A short time after your station receives the connect acknowledgement from the BBS station, an opening message appears. The opening message usually gives the station's call, the SYSOP's name, and the station's location. At the end of the opening message a prompt appears which usually includes your call, the date and time, and sometimes an abbreviated list of available commands. The prompt always ends with a > sign.

There are many commands available that are used to control the BBS. All commands are composed of a single letter, although many can be expanded further by the addition of another delimiter character which restricts the range of the command. The commands can be broken down into groups by the BBS function they control.

Mail System

The first group of commands covered are used to interact with the message (or mail) system as it is the most heavily used part of the BBS. Messages are sent to specified users by callsign. The user's call is automatically entered as the originating, or FROM, station. A BBS station can be specified to forward the message to.

In the early days of the WORLI system, the number of packet users was low enough that it was feasible to keep track of where each user was located and which BBS he usually used. Thus mail addressed to that user would be forwarded to that BBS. However, with the growing number of users logging onto the systems, it is not possible to keep track of where all the users are located. Thus, some of the forwarding responsibility has been transferred to the users by requiring them to include the address of the BBS station the recipient uses if it is not the same as the one which the message originated.

The first mail command most new users like to try is L(ist). The L command instructs the BBS to send the user a listing of all the new messages added since he last logged onto the system. The format of the listing is as follows: message number, message status, message size. TO call, FROM call, BBS call, date sent, and message title. (See Fig. 6-14.)

Msg#	TR	Size	To	From	@ BBS	Date	Title
125	Y	136	K3XXX	KBXXX		860509	Listing Msgs
124	Y	165	K3XXX	KA2XXX		860509	TNC for sale
120	Y	241	K3XXX	KB3XX		860507	monitor for sale
119	N	254	W3XXX	KB2XX		860507	Need help
116	N	389	ALL	W3XXX		860507	HAMFEST

Fig. 6-14. A sample message listing as would be received from a W0RLI-compatible BBS.

All messages are numbered sequentially, and that number is the message number. The message status indicates the message type and whether it has been read by the recipient. There are several types of messages including bulletins, private messages, ARRL messages, and traffic.

Bulletins are messages meant to be read by all who log onto the system. These include items for sale, upcoming events, system changes, and other news. Bulletins are indicated by message type B. Private messages are meant to be read only by the sender and recipient. They do not show up on a message listing to anyone other than the sender, the recipient, and the SYSOP; however, when sending or receiving private messages, they may be monitored by other users. Private messages are indicated by a message type P. ARRL bulletins and messages are indicated by a message type A. Traffic includes any messages meant to be forwarded by the *National Traffic System (NTS)*. They are indicated by message type T and are usually read by a local traffic operator and then entered into the traffic net on some other mode as packet is not yet fully utilized by the NTS. Regular messages may be listed and read by all users. The message type for regular messages is left blank.

The indicator as to whether the message has been read by the intended recipient shows an "N" if it has not been read by the intended recipient and a "Y" if it has. The indicator does not change if someone other than the intended recipient has read the message.

The message size consists of a number indicating the length of the message in bytes. If a message is very long (2500 bytes or more) it may be better to read it later if the channel is crowded or someone else is waiting to use the BBS. The FROM call indicates the station that stored the message on the BBS. The FROM call is usually the same as that of the user who sent the message; however, if the message was forwarded to another BBS, the FROM call shows the call of the BBS that last forwarded the message.

The TO call indicates the callsign of the intended recipient of the message. The TO call never changes regardless of forwarding. The BBS call is the callsign of the WRLI-compatible BBS that the message is to be forwarded to. The date gives the date that the message was stored on the BBS; if the date has changed when a message is forwarded, the new date is shown. The message title is entered by the user who sent the message. The title may change if the message is forwarded to indicate that the message was forwarded.

The L command also accepts several delimiters and arguments to allow users to selectively list messages. As mentioned earlier, sending the BBS a single L at the prompt results in a listing of all messages stored on the BBS (except private messages that are not from or to the user) since the user last logged on. The addition of a B after the L restricts the listing to all the bulletin type messages stored on the system (i.e., LB—List Bulletins). By typing an M rather than a B after the L, the system will list all the messages either sent by or for him (i.e., LM—List Mine).

The remainder of the list commands require the addition of an argument after the one or two letter command. The addition of a second L, a space, and a number (LL 4), causes the system to send a listing of the last specified number messages (except private not to or from the user) starting at the latest message and working backwards. For example, the command LL 5 (List Last 5) causes the BBS to send a list of five messages starting with the last message number and ending with the last message number minus 5.

The L command, a space, and a message number causes the BBS to send a listing of all messages stored on the BBS starting with the message number specified up to the last message number (L 1000). The L command with a < character, a space, and a callsign causes the BBS to send a listing of all messages from the specified callsign (i.e., L< KR3T); the private message restriction applies here and in the following examples also. The L with a > character, a space, and a callsign causes the BBS to send a listing of all messages addressed to the specified callsign (i.e., L> KR3T). The L command with an @ character, a space, and a BBS callsign causes the BBS to send a listing of all messages with the specified BBS callsign entered by the sender for forwarding (L@ WORLI).

The eight list commands listed above (L, LL #, L , LB, LM, L< call, L> call, and L@ BBS call) are very useful, and being familiar with them can help reduce the time needed to find a particular message out of the several hundred normally stored on a BBS. Notice that the command and delimiter are not separated by a space and that a space is required after the command (and optional delimiter) before an argument.

The next mail command to be discussed is the read command. The read command, R, is used to read messages. There are only two forms of the R command; R #, and RM. The R, a space, and a message number causes the BBS to send the entire text of the specified message number. It is not possible to read a private message not to or from the user or a message that does not exist. If this is attempted, the BBS responds with an appropriate message and redisplays the prompt. The RM command stands for Read Mine and causes the BBS to send the user the full text of all his messages that have not yet been read.

The third mail command discussed is the send command. The send command, S, is used to instruct the BBS to store a message. There are several versions of the send command. The first method is to send the BBS an S. This causes the BBS to prompt for the destination callsign and the message title. Once these are sent, the message can be transmitted. Once the message has been sent, a CTRL-Z is used to indicate that the message is over. The system then saves the message to disk, adds it to the message list, and redisplays the prompt.

An easier method is to include the addressee and optional BBS call with the S command (S KR3T or S KR3T @ WA3XXX). The BBS does not ask for the destination callsign if this method is used. The S command can include delimiters to indicate the type of the message (i.e., SP = Send Private, SB = Send Bulletin).

The fourth and last mail command, K, kills a message. The K command must have a message number as an argument (i.e., K 1203). If a K command is sent and the user's call is in the TO or FROM field, the message is deleted. However, if the user's call is not in the TO or FROM field of the message indicated by the message number argument, the message is not deleted, and a message informing the user that he cannot kill the message is sent by the BBS followed by the prompt. The K command followed by the M delimiter instructs the BBS to delete all messages addressed to the user (KM = Kill Mine). It is considered good practice to kill a message after it has been received if it is not of general interest.

File System

The next group of commands deals with the file system. The file system allows users to upload and download various files from the BBS. A file can consist of a list of local hamfests, a computer program, the minutes from the last packet meeting, a list of new packet operators, and almost anything else that can be represented in ASCII characters. Files are stored in a separate section of the BBS from messages. However, the commands are somewhat similar.

All files are stored on disk drives on the BBS computer. Each file is assigned a name when it is stored on the BBS. This name is referred to as the *filename*. If you are familiar with computers, you know that filename syntaxes vary from one system to another. The filenames for a TRS-80 operating under TRSDOS differ from those for an Apple under ProDOS. The filename syntax is controlled by the type of disk operating system the computer is utilizing. As you may recall from Chapter 1, the *Disk Operating System (DOS)* handles all accesses by the computer to information stored on the disk drives.

The W0RLI BBS program runs under the CP/M operating system and uses the same filename syntax as standard CP/M systems. In CP/M each disk drive is assigned an alphabetical designator ("A" for the first drive, "B" for the second, and so on). The individual files on a particular disk are stored on the disk's directory. Thus, obtaining a listing of the files stored on a disk is called "getting a directory."

The filenames under CP/M are composed of up to eight characters, a period, and a three character extension (for example, HAMFESTS.MAY). The filename usually describes the contents of the file.

An analogy can be drawn between a multi-drawer file cabinet and the file structure on a BBS. Just as a file cabinet's drawers each contain different file folders, each disk contains different files. On each file folder in a drawer is written a short, descriptive heading describing the contents of the file folder. Each disk contains a list of short, descriptive headings (filenames) describing the contents of the file. Finally, in the file folder is the actual file; just as the actual file is stored on the disk. Each drawer has a limited capacity; only so many sheets of paper can be put in a single drawer. On a disk, there is also limited capacity for information storage.

To locate and retrieve a file in a file cabinet, the drawer and filename must be known. If these are not known, it is possible to locate a particular file by looking at a list of each file folder in a particular drawer. The same can be done on a disk with the directory. What a directory actually does is access each drive and list the filenames of all the files stored on the disk in the drive; the size of each file and the total amount of storage space left on the disk is also usually given.

The W0RLI command to get a directory of the files stored on the BBS is W (for What's available). The W command alone causes the BBS to get a directory of each disk and send a list of the files on each disk along with their sizes to the user. The total amount of space consumed and the total amount of space remaining on each disk is also sent. It is also possible to get a directory of a specified drive. Supposing a BBS has three drives available for files labeled A, B, and C; a user could request a directory for each of the drives individually by sending the BBS the command W A:, W B:, and W C: respectively. It is also possible to get a limited directory through the use of *wildcard characters*.

A wildcard character is a special character that is recognized by the BBS in place of one of the components of the filename. The wildcard in CP/M is an asterisk (*). For example, suppose a BBS keeps an updated list of activities in various monthly files. All the files have the extension of the current month; however, the first eight characters or less before the period differ. To get a directory of all the updated files for the month of May, the user sends the BBS the command "W*.MAY." Thus, all files with the extension of MAY are listed.

Now that the files stored on the BBS are known, it is time to download a file. The command to download a file from the BBS to the user's station is D followed by a space and the filename (D HAMFESTS.MAY). The drive designator may be optionally included (D A:HAMFESTS.MAY); however, the BBS will search all drives for the specified file automatically. The download consists of straight ASCII text with no error-checking beyond that which is provided by packet itself. The wildcard character may be used here also, but be careful not to download too many files at once.

It is also possible to upload files to the BBS. The command to upload a file from a user's station to the BBS is U followed by a space, the drive designator, and the filename with extension. Be sure to select a filename not currently in use. One other very important point is to make sure the disk has enough space to save the file with room to spare. It is possible to disable (crash) the BBS if a file upload is attempted without adequate disk storage space. Once the BBS receives the command, it prompts for uploading to begin. After you finish sending a file, input a CTRL-Z to close the file.

There is no kill command for files; only the Sysop may delete files. The range of files available for downloading is great. The ARRL *Gateway* newsletter and the W5YI report are common files along with local maps that when printed out give a rough map of digipeaters, BBSs, and users in a specified area. Other files commonly available are listings of hamfest dates, tutorial files for new users, equipment modification instructions, computer programs, and lists of local users.

Gateway System

The next BBS area we'll discuss is gateway operations. A BBS gateway is usually a BBS with two TNCs, one on each serial port of the Xerox 820. Each TNC is connected to a radio operating on a different frequency, often on another band. A user accessing the BBS on one frequency and TNC can instruct the BBS to forward his or her transmissions to the other TNC and transmit on the other frequency. The command to access the gateway functions on a BBS is G. When the BBS receives the G command, it presents the user with a new gateway prompt. The commands available on the gateway prompt are C, R, U, and X. C is the command used to initiate a connection through the other TNC. The C command is followed by a space and a callsign. It is very similar to the TAPR connect command. When the C command and argument are received by the BBS when in the gateway prompt, the BBS instructs the other TNC to send a connect request to the specified callsign. The U command is used to send unnumbered information frames through the other TNC. It allows the user to broadcast data from his station, through the BBS, and onto the other frequency. The R command is used to return to the regular BBS prompt.

Other Commands

The remainder of the BBS commands are discussed below.

The J command lists all the stations heard by or connected to the BBS along with times that their transmissions were received.

This can be useful for monitoring activity or meeting new users. The T command is used to page the Sysop by ringing a bell at the BBS. If the Sysop is nearby, he or she will turn off the BBS, and QSO with the user directly.

The N command is used to tell the BBS your name (N Jon). The BBS then knows your name for customized prompts.

The X command is used to obtain an extended menu which lists all the commands available along with brief explanations. The long menu can be changed back to the short prompt by sending the X command a second time. The BBS usually uses the short prompt upon connection. New users may be more comfortable using the long menu for the first few BBS sessions, but it should not take long to learn the commands.

The H command calls up a somewhat lengthy help file which explains the general rules of the BBS.

The I command causes the BBS to send a brief information file describing the equipment used at the BBS.

The last command used during a BBS session is the B command. B means Bye, and indicates to the BBS that the user is through using the system, causing the BBS to disconnect automatically and ready itself for the next connection.

There are some additional messages that the BBS may send during a session with a user. These include error messages "*** What" or "*** I don't understand" when the user sends an improper command, "*** K3XXX just tried to connect" when another station attempts to connect to the BBS while the user is connected, and "*** Standby, the Sysop wants to chat with you" when the Sysop is switching from the BBS mode to the direct QSO mode. These messages are configurable by the Sysop, so they vary from system to system. The BBS also sends beacons containing the calls of the stations with new mail waiting for them.

NET/ROM Operations

In February 1987, amateur packet operators were first introduced to a new networking system known as NET/ROM which has since become very popular around the country. NET/ROM was briefly covered in Chapter 4. This section provides information on the operational aspects of NET/ROM. NET/ROM was developed by Ron Raikes WA8DED, and Mike Busch W6IXU, and is sold through their company, Software 2000.

NET/ROM, contained on a ROM chip that simply replaces the regular ROM chip in any TNC-2 compatible TNC, turns the TNC into a NET/ROM network node. Users can connect to other NET/ROM nodes, as well as conventional stations, with the benefit of point-to-point acknowledgements through the NET/ROM nodes. Each node maintains routing tables to other NET/ROM nodes, so it is not necessary to specify a path between nodes.

To use a NET/ROM node, just connect to it like a regular station. NET/ROM nodes are usually known by their alias as well as the station's callsign, and the alias is often a geographical or airport designator. You can connect using either designation. To locate a NET/ROM node in your area, ask other local packet operators or just monitor the local 2-meter operating frequency(ies).

Once you have connected to a NET/ROM node, you can obtain a listing of other NET/ROM nodes that the node can connect to by sending the NODES command. Simply type NODES and press the RETURN key. A list of other NET/ROM nodes will be displayed. If a node has an alias, it will be displayed immediately before the node's callsign.

You can now link up to any of the displayed nodes by using the CONNECT command. The NET/ROM CONNECT command is not the same as the TAPR user interface's CONNECT command. Keep in mind that you are already connected to a local NET/ROM node. You are simply sending a command to that node. For example, if one of the displayed NET/ROM nodes was W6IXU-1, you would type CONNECT W6IXU-1 and press the RETURN key. Your local NET/ROM node that you first connected to will now try to connect with W6IXU-1. If successful, you will receive a message indicating that you are connected to W6IXU-1. You can now use the NODES command again to see which NET/ROM nodes W6IXU-1 can link to.

Eventually, you will want to connect with an individual packet station or bulletin board. To do this, you use the CONNECT command again, only this time substitute the callsign or alias of the station you wish to connect with. In most areas of the country, it is futile to link through more than three NET/ROM nodes. Keep in mind that there may be several digipeaters between each node, so your actual path might be quite extensive.

OPERATING TIPS

At times, operating packet radio can be confusing to the beginner. With all the technical detail of establishing and maintaining a connection, some of the more routine tasks such as actually conducting a QSO can be overlooked. Also, there are many things that the beginning operator can do unknowingly to increase the congestion on a channel. The following operating tips are things that I have learned, along with comments from the many experienced operators with whom I have spoken.

The first tip seems obvious: don't send, or cause to be sent, any long files during times of heavy usage. This only results in tying up the frequency from others users. It is much better to switch to a less congested channel, if one is available, or wait until the channel activity has decreased to an acceptable level. Early morning and midday are general periods of low activity.

Keep in mind what was said about digipeaters earlier in the chapter. Use as few as possible to reach your destination. Again, if there is heavy channel activity, refrain from operating, or switch to another less used channel.

If you should connect to another user and get no immediate response, it is possible that he or she just didn't notice your connection. Try sending a few CTRL-Gs, which will cause terminals so equipped to ring a bell or buzz a buzzer. Some areas discourage this practice because the terminals of stations monitoring your transmissions or the digipeaters you are using may also ring. If that does not arouse the operator's attention, leave a message with your call, the path, and the time and date (KR3T via KR3T-1 at 14:00 12 JUL 85); don't forget to disconnect from his station. This way the operator will know who connected, when the connection took place, and how to link back to you (assuming the station is dumping connections to a printer or receive buffer).

Beacons are another area of concern. Use them as little as possible. Beacons should be as short as possible and carry only necessary information. In some areas of the United States, there are so many beacons being transmitted that they are colliding with each other (automatic channel congestion)!

A large amount of the packet activity centers around BBSs. When you first connect to a BBS, leave the Sysop a message with your name, location, and packet equipment. Most Sysops are interested in knowing who is using their systems, and they are good people to know.

When sending commands to a BBS, do not repeat the command if nothing happens immediately. This is difficult for many RTTY operators because they assume the first transmission was lost. However, on packet, the command will get through or the user will retry out and be disconnected.

It is a good idea not to stay connected to a BBS for a long time during prime time. Also, do not download a large file if there is moderate activity on the channel. Before you connect to a BBS for the first time, make sure you have a reliable link by connecting to yourself through it (KR3T = me, KR3T-3 = BBS: C KR3T V KR3T-3). Otherwise, you may time out the BBS.

A time out on a BBS is different from a time out on an FM repeater. If the BBS does not hear from your station for a period of time (usually four minutes) you are disconnected. What usually happens is the link to the BBS fails and your station retries out and disconnects. Meanwhile, the BBS is still sending you frames until it retries out or waits for four minutes for your next command.

When operating HF packet, tuning is critical. Because packet transmissions are short, it can be difficult to tune them in. A tuning indicator is a great help in this area. Also, HF packet is not nearly as fast as VHF because of the lower baud rate used and the problems with interference and propagation associated with HF operation.

I hope these suggestions make it easier for you to operate packet radio. Most packet operators are very understanding of mistakes and are more than willing to help out newcomers. Most of these suggestions have to do with conserving frequency space. This is because at present, most of the packet operation is done on one or two frequencies using digipeaters. Once backbone links on other frequencies or true level 3 network nodes develop, the problems with channel congestion will not be as serious and detrimental to packet activity. Even with the present congestion problems, packet is a wonderful mode to operate; and with the future developments already being worked on by many amateurs worldwide, it will become even better.

CONCLUSION

This chapter has run the full course from setting initial parameters to connecting to bulletin board systems and transferring files to general operating tips. It is my hope that having some prior knowledge of the techniques used in operating packet radio will take away some of the nervousness and concern from operating packet and replace it with understanding and excitement. If things don't seem to be going right at first, relax, sit down, and think about exactly what you are doing. Sketching out the steps you are taking and the paths you are using may be helpful in diagnosing your problem. Most new users quickly adapt to packet radio and it rapidly becomes one of their favorite modes.

The next chapter takes a look at the packet radio equipment and accessories that are available now, and those to come in the near future.

Chapter 7

Equipment and Accessories

This chapter provides a comprehensive overview of currently available amateur packet radio equipment. Some of the equipment listed, is no longer manufactured, but is included because it represents a major development in amateur packet radio. You may also run across this equipment in the used market or when talking with other operators. For these reason, it is a good idea to have knowledge of the various products that are in use today.

Previews of packet products currently under development were provided by various manufacturers. They are included in this chapter to give you an idea of what to expect in the future. Be sure to check with the manufacturer as to the progress of their product's development.

This chapter includes TNCs, software, and hardware additions. I would like to take this opportunity to thank those companies and groups that were kind enough to provide me with equipment loans and/or information: AEA, DRSI, GLB, HEATHKIT, Kantronics, MFJ, Pac-Comm, and VADCG.

The products are listed in alphabetical order by manufacturer, group, or individual; the addresses are included in Appendix C. Any pricing information given is subject to change.

AEA
PK-232

The AEA PK-232 is a very popular multimode data controller. (See Fig. 7-1.) It is capable of operating on CW, Baudot RTTY,

Fig. 7-1. The AEA PK-232. (Photo courtesy of AEA.)

ASCII RTTY, AMTOR, Packet, and Facsimile. The PK-232 has twenty-one front-panel indicators. Two radio ports allow for the simultaneous interfacing of an HF and a VHF transceiver. The list price of the PK-232 is $319.95.

AEA also sells two terminal programs for use with the PK-232. PC Pakratt with FAX is for the IBM-PC and compatibles. Com Pakratt with FAX is a ROM cartridge for the Commodore 64 and 128. Both programs have a split-screen display and can transmit and receive facsimile.

PK-88

The PK-88 is a standard TNC-2-like TNC. (See Fig. 7-2.) It features a "maildrop" through which others can exchange information. The PK-88 is compatible with the TCP/IP protocol and can be modified for NET/ROM operation. The PK-88 lists for $119.95.

Fig. 7-2. The AEA PK-88. (Photo courtesy of AEA.)

PM-1

The AEA PM-1 (Packet Modem) is designed to interface between your existing TNC and radio. (See Fig. 7-3.) The PM-1 contains independent dual-channel filtering for maximum sensitivity and selectivity under poor HF conditions. The PM-1 is optimized for 300 baud operation; a frequency shift of 200 Hz or 600 Hz may be selected from the front panel. A bar graph tuning indicator is provided to assist with precise tuning. List price is $199.95.

RFM-220 Radio Modem

The RFM-220 Radio Modem is a high-speed packet modem and 220 MHz transceiver. When interfaced to the external modem connector on most standard TNCs, the RFM-220 can be used to send and receive data at 19,200 baud. The RFM-220 can also be used as a channelized 220 MHz FM voice radio. The RFM-220 lists for $995.95.

PKT-1

The PKT-1 was the first TAPR TNC-1 clone. However, it differs in many respects from the original TAPR design. These differences include no parallel port (one is available as an option), no wire wrap area, 12 Vdc operation, and a 5-pin radio connector with redundant audio and AFSK outputs. The PKT-1 uses the full TAPR TNC-1 command set and is operationally compatible with the TAPR TNC-1. List price was $499.00. (See Fig. 7-4.)

Fig. 7-3. The AEA HF Modem, PM-1. (Photo courtesy of AEA.)

AEA 145

Fig. 7-4. The AEA PKT-1. (Photo courtesy of AEA.)

The PKT-1 is virtually identical to the TNC-1. There are plated holes on the printed circuit board for the mounting of the parallel port components. The user interface is almost identical to the TAPR TNC-1's; however, all references to TAPR have been replaced with AEA. The owner's manual is not nearly as extensive as the one included with the TAPR TNC-1, and a second technical manual is available as an option from AEA. The PKT-1 is no longer manufactured.

Pakratt 64

The PK-64 is an excellent example of a machine-specific TNC. It connects to the cartridge port of a Commodore 64 or 128. This unit not only operates packet but also CW, RTTY, and AMTOR. For HF work, the optional HFM-64 (High Frequency Modem) is recommended. The HFM-64 installs inside the PK-64 and replaces the internal PLL modem for HF operation. The HFM-64 adds a tuning indicator and threshold control to the front panel of the PK-64. (See Fig. 7-5.)

There is no software necessary in the Commodore; all software is included in the PK-64. The user interface of the PK-64 utilizes

146 Equipment and Accessories

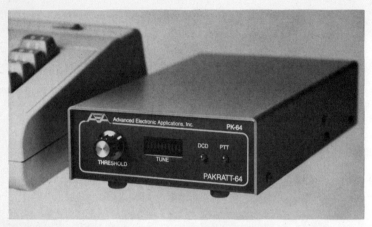

Fig. 7-5. The AEA PK-64. (Photo courtesy of AEA.)

function keys and parameter screens. It also includes a download buffer and tuning indicator and uses the TAPR command set. Like the PKT-1, the PK-64 includes a 5-pin radio connector and redundant audio plugs. The list price was $219.95 for the basic PK-64. The PK-64 is no longer manufactured.

BILL ASHBY AND SON

PAC/NET Board

The PAC/NET is a TNC board similar in design to the VADCG TNC. It does not include a modem and utilizes its own user interface. Prices range from $80 to $240 depending on configuration.

DRSI

PC∗Packet Adapter

The DRSI PC∗Packet Adapter is a dual-port packet radio TNC that plugs into an IBM-PC or compatible computer. (See Fig. 7-6.) The PC∗Packet Adapter has an on-board 1200 baud modem for VHF operation. A second output, which can be configured for either RS-232 or TTL signals, can be interfaced to an external modem, such as the HF∗Modem. The PC∗Packet Adapter lists for $139.95.

Fig. 7-6. The DRSI PC∗Packet card. (Photo courtesy of DRSI.)

HF∗Modem

The DRSI HF∗Modem is a 300 baud modem and tuning indicator that can be interfaced to DRSI's PC∗Packet Adapter for HF operation. (See Fig. 7-7.) The HF∗Modem can also be used with any standard TNC-2 via a modem disconnect adapter available from DRSI. The HF∗Modem lists for $79.95.

GLB

PK-1

The GLB PK-1 was the first commercially produced TNC, introduced in August 1983. The PK-1 does not include any status indicators other than a power on indicator. The PK-1 implements HDLC in software, so when a frame is being disassembled the processor is running at close to full capacity. Therefore, when the unit is accepting input from the user via the RS-232 port it cannot disassemble incoming frames. The PK-1 comes in a variety of configurations. List prices range from $109.95 to $169.95.

148 Equipment and Accessories

Fig. 7-7. The DRSI HF*Modem. (Photo courtesy of DRSI.)

PK-1L

The PK-1L is a portable, low-power version of the PK-1. It draws about 25 mA @ 12 Vdc making it ideal for portable operation. As in the case of the PK-1, the PK-1L does not have any status indicators (not even a power on indicator). The unit can be powered for several hours from an ordinary 9 Vdc battery. (See Fig. 7-8.) List price is $209.95.

I used the PK-1L in a portable station for several weeks. (See Fig. 7-9.) I quickly adjusted to the GLB command set and began

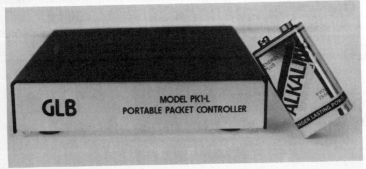

Fig. 7-8. The GLB PK-1L. (Photo courtesy of GLB.)

Fig. 7-9. A portable packet station incorporating the PK-1L. The perfboard to the right of the Model 100 is a PTT controller for the IC-02AT.

to appreciate its compactness and efficiency. The PK-1L went on two trips with me during preparation of the first edition of this book (the first was to Georgia, and the second was to the 1986 TAPR annual meeting in Tucson, Arizona). The PK-1L would make an excellent low-power AX.25 digipeater for use at remote locations.

Netlink 220

The GLB Netlink 220 is a high-speed modem operating on the 220 MHz band. (See Fig. 7-10.) The Netlink 220 can be interfaced

Fig. 7-10. The GLB Netlink 220. (Photo courtesy of GLB.)

to any standard TNC-2 via the external modem connector. Transmission speed is 19,200 baud and turnaround time (from transmit to receive and back) is about 1 millisecond. The Netlink 220 lists for $699.95.

TNC-2A

The TNC-2A is another TAPR TNC-2 clone. The only difference between the TAPR TNC-2 and the GLB TNC-2A is that GLB has replaced the cylindrical RAM backup battery with a flat button cell and a slightly different cabinet. The TNC-2A came only in kit form. List price was $169.95. The TNC-2A is no longer manufactured.

HAL

RPC-2000

The HAL RPC-2000 is a dual-port packet radio TNC contained on a full length IBM-PC or compatible card. A VHF modem is included on-board, and an external modem, such as the ST-7000, can be interfaced to the second port. Contact HAL for pricing information.

ST-7000

The HAL ST-7000 HF-Packet Modem is designed specifically for 300 baud HF packet operation. The ST-7000 contains a built-in tuning indicator and is fully compatible with all existing TNCs. Contact HAL for pricing information.

HAMILTON AREA PACKET NETWORK

HAPN Packet Adapter

The HAPN packet adapter is a plug-in TNC for IBM-PC and compatible microcomputers. The HAPN packet adapter plugs directly into one of the slots of the PC and contains a built-in modem which interfaces to a radio via a DB-9 connector. The HAPN packet adapter is capable of running AX.25, V-1, and V-2. List cost is $199.00 for an assembled and tested board with AX.25 software.

HEATHKIT

HK-232

The Heathkit HK-232 PackKit is Heathkit's licensed kit version of AEA's PK-232. The HK-232 is electrically identical to the PK-232,

Heathkit 151

Fig. 7-11. The Heathkit HK-232. (Photo courtesy of Heathkit.)

and its case, while following the same layout, is cosmetically different from the PK-232. (See Fig. 7-11.) The HK-232 in kit form lists for $279.95; a 12V power supply is $19.95; and an optional technical manual is $24.95.

HK-21

The Heathkit HK-21 is an interesting device—a pocket-sized TNC. (See Fig. 7-12.) It is ideal for all VHF packet portable situations. The HK-21 is TNC-2 software compatible and also includes a built-in mailbox system. It is smaller than a pack of cigarettes, and an optional battery pack can be installed inside the unit, eliminating the need for an external power supply. A regular

Fig. 7-12. The Heathkit HK-21. (Photo courtesy of Heathkit.)

Equipment and Accessories

RS-232 port is provided for interfacing to almost any terminal, including portable computers. Two radio interface methods are provided: direct mic and speaker cables for most HTs (such as the IC-02AT) with full PTT control, and a conventional port for interfacing with other transceivers. The HK-21 lists for $219.95 and the optional internal battery pack is $17.95.

HD-4040

The HD-4040 was another TAPR TNC-1 clone and came in kit form. (See Fig. 4-13.) It was identical to the TAPR TNC-1 in all respects. The list price was approximately $250.00. The HD-4040 is no longer manufactured.

KANTRONICS

KAM

The Kantronics All Mode (KAM) is Kantronics' answer to AEA's PK-232. (See Fig. 7-14.) The KAM operates on the following modes: Facsimile, CW, RTTY (both ASCII and Baudot), VHF and HF packet, and AMTOR. A built-in tuning indicator and dual-ports for the simultaneous connection of a VHF and an HF radio are provided. Kantronics also markets proprietary terminal software for use with the KAM (as well as most of their other TNCs). Kanterm 64 and 128 is available for the Commodore computers, and Pacterm is available for the IBM-PC and compatibles. MAXFAX, a program for reception and storage of facsimile images, is available for both computer systems. The KAM lists for $319.00, Kanterm and Pacterm are each $29.95, and MAXFAX is $19.95.

Fig. 7-13. The Heathkit HD-4040.

Kantronics 153

Fig. 7-14. The Kantronics KAM (Kantronics All Mode). (Photo courtesy of Kantronics.)

KPC-2

The Kantronics Packet Communicator-2 (KPC-2) is a revised model of their original KPC. (See Fig. 7-15.) It is somewhat similar to the TAPR TNC-2 and uses an almost identical user interface. List price is $219.95.

The KPC-2 modem section is easily switched between VHF and HF tones; however, no tuning indicator is provided. The KPC-2 runs on 12 Vdc. The RS-232 connector does have two non-standard pins which may interfere with your terminal if they are connected.

KPC-4

The Kantronics Kantronics Packet Communicator-4 (KPC-4) is a true dual-port TNC with two simultaneously operable radio ports. Automatic gateway operation between ports is provided. Crossband

Fig. 7-15. The Kantronics KPC-2. (Photo courtesy of Kantronics.)

154   Equipment and Accessories

Fig. 7-16. The Kantronics KPC-2400. (Photo courtesy of Kantronics.)

and in-band split gateway operation is possible. Or, two independent packet connections can be made, one on each port. The KPC-4 lists for $329.00.

KPC-2400

The Kantronics Packet Communicator-2400 (KPC-2400) includes the 300 baud and 1200 baud KPC-2 modem in addition to a 2400 BPS PSK modem. The PSK modem operates at 1200 baud with a signaling rate of 2400 BPS (See Fig. 7-16.) List price is $329.00.

2400 TNC Modem

Kantronics is making available external 2400 BPS modem cards for the TAPR TNC-1 and TNC-2 and their clones. These modem cards attach to the external modem connector on the TNCs and allow them to communicate with their KPC-2400 as well as other similarly-equipped TNCs at 2400 BPS. List price is $149.00.

MFJ

1270B

The MFJ 1270B is MFJ's latest version of their 1270 TNC-2 clone. (See Fig. 7-17.) The 1270B features TNC-2 compatible firmware, and lists for $139.95.

Fig. 7-17. The MFJ 1274 (upper) and MFJ 1270B (bottom). (Photo courtesy of MFJ.)

1274

The MFJ 1274 is a VHF/HF TNC with a built-in tuning indicator. (See Fig. 7-17.) The 1274 is basically a 1270B with an HF modem and tuning indicator. It lists for $169.95.

1278

The MFJ 1278 is MFJ's multi-mode data controller. (See Fig. 7-18.) Similar to the AEA PK-232 and Kantronics KAM, the 1278 features operation on HF and VHF packet, Baudot and ASCII RTTY, CW, facsimile, and Slow Scan Television (SSTV). The 1278 also has a built-in contest memory keyer. The Slow Scan and FAX modes allow for both transmission and reception of images. The 1278 lists for $249.95.

1273

The MFJ-1273 is a TAPR compatible tuning indicator which plugs into the TAPR tuning indicator connector on most TNC-1s, TNC-2s, and clones. The 1273 lists for $49.95.

MICROLOG

ART-1

The Microlog ART-1 is an all-mode digital interface for the Commodore 64 and 128 microcomputers. It initially comes equipped for CW, RTTY, and AMTOR operation, and packet capability is

156 Equipment and Accessories

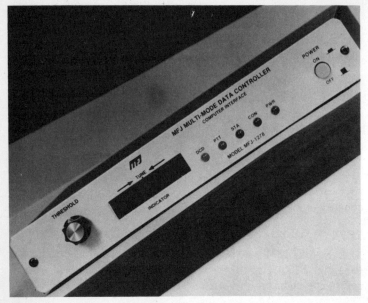

Fig. 7-18. The MFJ 1278. (Photo courtesy of MFJ.)

available as an option. The ART-1 packet option uses either TAPR compatible commands or unique Microlog commands. List price is $199.00.

PAC-COMM

TNC-220

The Pac-Comm TNC-220 is designed to be a successor to the TNC-200 and other TAPR TNC-2 clones. It features a single-chip modem section that is software configurable between two radio ports. The standard configuration supports one HF and one VHF radio port, each with its own radio cable connector. Switching between ports is done entirely in software, so no cable switching or retuning is required. The modem includes a bypass header for addition of an external modem. The TNC-220 uses the TAPR command set and supports AX.25.

TINY-2

Pac-Comm's TINY-2 is an improved version of the TAPR TNC-2. (See Fig. 7-19.) The TINY-2 comes with TAPR firmware and is NET/ROM compatible. The TINY-2 lists for $119.95.

MICROPOWER-2

The Pac-Comm MICROPOWER-2 is a low-power version of the TINY-2, consuming about 40 mA at 9 to 12 Vdc. (See Fig. 7-20.) The MICROPOWER-2 lists for $159.95.

TNC-200

The Pac-Comm TNC-200 is another TAPR TNC-2 clone. (See Fig. 7-21.) The link and terminal baud rates are set via DIP switch-

Fig. 7-19. The Pac-Comm TINY-2. (Photo courtesy of Pac-Comm.)

Fig. 7-20. The Pac-Comm MICROPOWER-2. (Photo courtesy of Pac-Comm.)

158 Equipment and Accessories

Fig. 7-21. The Pac-Comm TNC-200. (Photo courtesy of Pac-Comm.)

es on the back panel. The modem configuration can be changed by altering components mounted on a DIP header (a device which allows for the mounting of discrete components and plugs into an IC socket). List prices range from $39.95 for a bare board and manual to $219.95 for an assembled and tested CMOS version with power supply.

The TNC-200 comes in a variety of configurations both kit and assembled. It runs on 12 Vdc. The radio port is a 5-pin DIN socket. I have found the TNC-200 to be a good, reliable TNC-2 clone.

PC-120

The PC-120 is a TNC on a half-size IBM-PC compatible card. It includes TNC software and built in modems along with RS-232 interfaces for external modems. (See Fig. 7-22.)

Surface-Mount TNC

Pac-Comm is currently working on an ultra-miniature surface-mount TNC. (See Fig. 7-23.) This TNC will be small enough to be installed inside the case of some HTs. Pac-Comm already markets miniature internal data modems. The Pac-Comm HT-1200 is a baud modem with a TTL digital interface. The Pac-Comm HT-232 is an RS-232 level converter for the HT-1200. The HT-1200 lists for $31.75, and the HT-232 lists for $10.50.

DR-100/DR-200

The Pac-Comm DR-100 and DR-200 are packet radio switch/repeater modules which have been designed for dedicated

Fig. 7-22. The Pac-Comm PC-120 card. (Photo courtesy of Pac-Comm.)

Fig. 7-23. The Pac-Comm ultra-miniature TNC circuit board (upper left), the TNC's 1200 baud modem circuit board (upper right), the HT-1200 (lower left), a quarter (lower middle), and the HT-232 (lower right). (Photo courtesy of Pac-Comm.)

network support service. The DR-100 is a single-port device which functions as a low-cost single-frequency digital repeater. The DR-200 is a dual-port design; the DR-100 may be upgraded to dual-port capability. Both units are based on the Z-80 microprocessor with a capacity of 32K each of ROM and battery-backed RAM. Provisions are included for the addition of an RS-232 compatible terminal. The DR-100 and DR-200 are provided with user selected software.

MacPacket

MacPacket is a packet terminal program for the Apple Macintosh computer. It features a split screen (separates send and receive text), receive buffer, menus, and digipeater routing tables. List price is $49.95.

PC-PACKET

PC-PACKET is a packet terminal program for the IBM-PC and compatibles. It features a split screen, windows, function keys, and receive buffer. List price is $49.95.

RICHCRAFT ENGINEERING

Robert Richardson W4UCH has developed a packet radio software program for TRS-80 models I, III, and IV microcomputers. The only hardware necessary is a modem and radio. The software turns the computer into a functional TNC. There are two systems available: VADCG and AX.25. Volume 1 (a 232-page book discussing the process of implementing the VADCG protocol in software) cost $22.00, and a disk containing the software costs $29.00. Volume 2 (a 253-page book discussing AX.25) also costs $22.00, and the disk of software costs $29.00.

SOFTWARE 2000

NET/ROM

Software 2000's NET/ROM is a very popular level 3 networking system developed by Mike Busch W6IXU and Ron Raikes WA8DED. NET/ROM comes on an EPROM chip that simply replaces the ROM chip in TNC-2 compatible TNCs. See Chapter 4 for more information on NET/ROM operation. Contact Software 2000 for pricing and licensing information.

S. FINE SOFTWARE

Macket

Macket is a powerful packet telecommunications program for the Apple Macintosh and runs on the 512K, 512e, Mac Plus, Mac

SE, and Mac II computers. Extensive windowing, special host-mode support for many TNCs, and uploading and downloading are some of its many features. Macket will work with all TNCs that have an RS-232 port. Macket lists for $39.95.

TAPR

TNC-1

The TAPR TNC-1 is the *de facto* standard for all present hardware TNCs. The TAPR TNC-1 bridged the gap between the earlier hardcore packet radio experimenters and today's more user-oriented operators. The TAPR TNC-1 was the TNC that opened the world of packet radio to thousands of amateurs.

The TAPR TNC-1 uses a 6809 microprocessor and features LED status indicators, a parallel status port, and an RS-232 serial I/O port. The TNC-1 uses *Non-Volatile RAM (NOVRAM)* for the storage of permanent parameters such as callsign, transmitter delay, and retries. A wire wrap area is provided for modifications. The modem was designed with certain components mounted on DIP headers so that modem configuration could be changed by simply changing headers.

The software implements both AX.25 Version 1 and VADCG (V-1). Up to eight digipeaters are allowed. The user interface was developed to be as user-friendly as possible.

The manual is written with the first-time packet operator in mind, including a tutorial beginning which introduces new users to packet radio. Several technical sections allow the manual to serve as a technical reference for advanced operators. The assembly section is detailed and sufficient graphics are included to guide the user painlessly through the assembly process.

Although the TAPR TNC-1 is no longer produced by TAPR, its expansive user base and the proliferation of clones guarantee its continuing place in amateur packet radio.

TNC-2

The TNC-2 was developed by TAPR to overcome some of the inherent disadvantages of the TNC-1. While the TNC-1 was (and is) an excellent unit, it was rather expensive, large, and required

110 volts. TAPR sought a new TNC with a lower cost, 12 Vdc operation, smaller size, and lower weight. The end result of these efforts was the TNC-2.

The TNC-2 had a host of new features not available previously. The TNC-2 was the first TNC to fully implement the AX.25 Version 2 protocol. The monitor commands and capabilities were increased. A real-time clock allowed time stamping of incoming packets. All parameters were saved with battery-backed RAM.

The hardware of the TNC-2 also differed from that used in the TNC-1. A Z-80 microprocessor was chosen for the TNC-2 partially because of the large amount of software development systems available for it. The modem section remained virtually identical to the TNC-1's. The TNC required only a single 12 Vdc supply rather than access to 110 volts.

Like the TNC-1, the TNC-2 has been licensed and widely copied. The TNC-2 set a new standard for amateur packet radio TNCs.

NNC

The TAPR Network Node Controller (NNC) is the latest major project that TAPR has undertaken. Descriptions of the anticipated capabilities of the NNC and possible implementations are given in Chapter 4. What follows is a technical description of the NNC.

The NNC is composed of a microprocessor, four HDLC ports, two asynchronous ports, a parallel port, a real-time clock, and up to half-a-megabyte of memory. The microprocessor chosen is the 64180 (a high speed, high performance Z-80 like processor). Because it is Z-80 compatible, it can immediately run most software written for the Xerox 820 and TNC-2. The serial ports can be configured to act as RS-232 compatible serial interfaces for the addition of terminals and serial printers.

VADCG

TNC+

The VADCG TNC+ is a revised version of their original TNC. (See Fig. 7-24.) The TNC+ comes in kit form with parts for $168.00.

WA8DED

Ronald Raikes WA8DED has developed alternate software with a different command set and user interface for the TAPR TNC-1

Fig. 7-24. The VADCG TNC+.

and compatibles. His program fits on two EPROMs and plugs into the ROM sockets on the TNC-1 in place of the TAPR ROMs.

The WA8DED firmware differs from the TAPR system in several respects; the most obvious is the user interface structure. Where the TAPR system has three modes, the WA8DED system has only two (a user mode and a host mode for computer controlled operation). All text entered from the terminal is transmitted when the CARRIAGE RETURN (CR) key is pressed except when the text is preceded by an ESC (Escape) character. Thus, all commands are preceded by an ESC character.

The WA8DED firmware allows multiple connections and monitoring of all frame types. All commands are one- or two-letters, and whenever possible, they are modeled after the TAPR command abbreviations. The WA8DED software is in the public domain for non-commercial use and is available from a variety of sources. If you need help in obtaining a copy of the software, contact WA8DED.

CONCLUSION

This concludes this chapter on currently-available amateur packet radio products. There are bound to be many products that I have not covered. That does not mean that I have purposely excluded them for any particular reason; rather, because of space and time considerations, I am limited in the number of products I can include.

The next chapter is an overview of what can be expected for amateur packet radio in the future.

Chapter 8

The Future of Amateur Packet Radio

Amateur packet radio is a rapidly evolving communications mode. While there have been many advancements since the original amateur packet activity in Canada, there are still many more improvements being worked on. These changes will provide additional capabilities and expand the usefulness of packet. This chapter introduces areas of current development and areas which might see development in the future.

Many new areas of packet radio are being explored and developed. These include packet hardware, software and protocols, and operating practices. In the area of hardware, work is underway on developing high-speed modems, digital transceivers, integrated TNCs, and packet satellites. Future developments in software and protocols include improved software TNCs, more advanced user interfaces, and higher level protocols. Changes such as less VHF DX operation, growth of local networks, UHF and VHF gateways and level 3 networks will reduce packet congestion, and improved operating practices for packet activity can also be expected.

HIGH-SPEED MODEMS

Work on high-speed modems has been going on almost as long as packet radio has been in existence. The Bell 202 standard was chosen for amateur packet operation on the VHF and UHF bands because of the readily available components, the ease of interfacing with most radios, and their availability on the surplus market. The 1200 baud transmission speed is a big improvement over the 300 baud and lower speeds usually used in amateur digital communications. However, as packet grows, even 1200 baud is not fast enough.

With many users sharing the same channel, it is beneficial to minimize transmission time. Also, higher speeds reduce the time required for long file and message transfers. There are alternatives to 1200 baud Bell 202 modems being proposed, but developing and selecting one optimum for packet operation has not been easy.

Because VHF and UHF links are usually reliable and interference-free, better modulation techniques and much higher speeds are feasible. Limitations are imposed by FCC regulations and the TNCs and transceivers being used. By utilizing a modulation technique such as PSK in which more than one bit may be encoded per baud, high BPS rates can be obtained without a significant increase in bandwidth. Many transceivers may require modifications for use with high-speed modems. These modifications may include the removal or bypass of filters and other components designed to support voice transmission and reception. Since most amateurs will not want to modify their radios for packet use, modem systems which interface simply to the transceiver through the standard mic and audio connectors are most desirable.

Currently, work is being done on 9600 baud modems (direct FSK, 1 bit per baud) by several individuals and at least one manufacturer. Initially, these modems will be used for backbone links between BBSs for networking purposes. TAPR has produced a semi-kit of a 9600 baud modem developed by Steve Goode K9NG for experimental use. Manufacturers, such as AEA and GLB, have introduced commercial high-speed modems for packet use. At first these modems will be for network linking purposes only, but they may eventually filter down to individual users. The primary obstacle to this is the need for compatibility between modem systems and the transceiver modifications that may be necessary for the high-speed operation.

DIGITAL RADIOS

The problems associated with interfacing high-speed modems to regular voice grade transceivers will eventually be solved by the development of digital transceivers. Digital transceivers will be designed expressly for use with high-speed modems and, as such, their components will be carefully chosen to meet this objective. These radios will probably not be able to operate in the voice mode. They most likely will take the form of modular RF decks that will be incorporated into the modem design. Preliminary work is being done in the area of digital radios, and it will not be too long before they are in use in amateur packet radio.

Taking the concept of digital radios one step further by also including the TNC in the unit is another possibility. Interference problems may have to be resolved, but the advantages of reduced cables, a single power supply, and portability make the idea attractive. Incorporating a terminal will result in a complete packet station in a single package.

Amateur Packet Radio Satellites

Packet radio satellites promise to add many additional capabilities to packet radio. Worldwide networking is feasible via satellite, and there are several packet satellites planned for launch in the near future. Individual users may access the satellites directly or through shared ground-based gateway stations equipped to automatically track the satellites and relay all traffic sent to the gateway through the satellite selected. For more information on amateur satellites and operating conditions, consult *The Satellite Experimenter's Handbook* by Martin Davidoff listed in the bibliography, or contact AMSAT.

One of the packet satellites is the Phase III-C, now known as Orbital Satellite Carrying Amateur Radio 13 (OSCAR 13), which has launched into a highly elliptical orbit on 15 June 1988 from French Guiana. OSCAR 13 contains a device known as a RUDAK which serves as a space-borne digipeater. While RUDAK has not been activated yet, it should be operating in the near future. RUDAK, which was designed by the West German affiliate of AMSAT, contains memory for use as buffers and for future uploading of programs from the ground. The program space gives RUDAK the ability to adjust to new protocols or operating conditions. The uplink to RUDAK will be 2400 baud PSK, and the downlink will be 400 baud PSK. See Chapter 4 for more information on the various encoding techniques. Because of the modulation characteristics chosen, special modems will be needed for use by ground stations that want to communicate with RUDAK.

The next packet satellite is known as JAS-1. JAS-1, now known as FUGI OSCAR-12 (FO-12) or Fugi, was launched in early August

1986. JAS-1 was developed by JAMSAT, the Japanese affiliate of AMSAT, and features an on-board flying mailbox similar to terrestrial packet BBSs.

JAS-1 carries two separate Mode-J transponders (2-meter uplink/70-cm downlink). The JA transponder is a standard linear transponder similar to those on current amateur satellites. The JD transponder is a digital transponder featuring four input channels using Manchester coded FM. There is one downlink channel using PSK. The AX.25 version 2 protocol is used and the BPS rate is 1200.

Another packet satellite being developed currently is (PACket SATellite) (PACSAT) by AMSAT. This satellite is designed exclusively for packet operation and will feature a full-blown flying mailbox with over 1 megabyte of memory. At this point, many design variables are still open, so it is not possible to give specific information regarding modulation methods. Current plans call for PACSAT to be placed in a low, near polar orbit. Launch was to be via a Space Shuttle "get-away special"; however, because of the setback in the shuttle program and some funding problems, it is doubtful that PACSAT will be launched in the near future.

One area of concern regarding packet satellites is compatibility. So far, each satellite has different modulation characteristics. This requires ground stations to have separate modems for each satellite. TAPR has indicated an interest in providing modem kits for the satellites. It may be possible to have a single modem which by onboard changes can be reconfigured for compatibility with the various satellites.

SAREX 2

Another area of packet space communications is the proposed SAREX 2 packet radio experiment. SAREX stands for Shuttle Amateur Radio EXperiment. SAREX 1 was the SSTV experiment flown on Spacelab 1. Following the successful operation of FM and SSTV from the Space Shuttle, several packet operators began to formulate a plan to place an amateur packet station aboard the Space Shuttle. The National Aeronautics and Space Administration (NASA) was cooperative and, based on the success of the earlier amateur radio operations from the Space Shuttle, approved the packet experiment concept. Work has begun on the hardware and software necessary to operate packet and still meet NASA specifications. A TRS-80 Model 100 portable computer was donated to the project by Radio Shack and was "flight hardened" to withstand the stress of a space mission. TAPR agreed to supply a flight hardened TNC-2. A BBS was developed for the Model 100 along with special coding for the TNC-2. The TNC-2 and power supply were mounted in a shielded case attached to the bottom of the Model 100. The radio will be the same Motorola 2-meter FM handheld used on the previous amateur Shuttle operations.

The end result is an automated packet station for use on the Shuttle which will allow stations to connect to it, send their contact serial number, and initiate a disconnect. Rather than store the log of stations worked aboard the Shuttle, special beacons will periodically send the calls of the most recent stations heard by the Shuttle and those stations that actually connected, received a serial number, and disconnected. Utilizing the multiconnect feature of the TNC-2, up to 9 QSOs with the Shuttle can be in progress at any one time. A terminal mode is also included so that the astronaut may conduct live QSOs.

SAREX 2 has not flown yet because of some administration problems with NASA and has suffered a more permanent setback due to the shuttle tragedy. Now that the shuttle program is back on track, SAREX 2 may have an opportunity to fly in the foreseeable future.

SOFTWARE AND PROTOCOLS

While the future plans for packet hardware are impressive, an equal amount of work is going into the software, protocols, and networks necessary to make the hardware function. These nonhardware developments can take many forms; some are as simple as a 10-line BASIC terminal program or as complex as the coding for a new TNC or a high-level protocol.

Software TNCs

Software TNCs will probably never achieve widespread use because of the disadvantages discussed in Chapter 5. However, there is a potential for limited use if a full-featured program was designed for a popular computer system. The protocol specifications are easily obtained, so anyone with a good knowledge of programming and a great deal of time could develop such a program.

User Interfaces

User interfaces are another area of programming that have seen a great deal of effort. As more nontechnical amateurs enter the ranks of packet operators, the need for increasingly user-friendly interfaces will develop. While long lists of sometimes cryptic commands are now necessary to allow complete control of the packet station, many users will want more informative commands within the user interface than, for example, a "CMD:" prompt. Menus, help screens, and tutorials would be a great help to many users attempting to learn the system.

A new generation of user interfaces should be developed which will meet the needs of the casual packet operator. In the case of machine-specific or software TNCs, the advanced user interfaces could be programmed in along with the rest of the command set. However, in universal hardware TNCs whose primary objective is compatibility between different systems, a higher level user interface will have to be developed. This interface software will not only serve as a terminal emulator, but also handle all communications with the TNC. The user will not see the individual commands or the "CMD:" prompt. Rather, the user will be presented with a series of "windows," menus, or whatever fits best with the specific computer system being used as a terminal.

For example, to initiate a connect, the user would simply position the cursor over "CONNECT" on the menu and press RETURN. The user could then be prompted to enter the appropriate calls and relay stations if necessary. To change a parameter, the user could select another menu and then select the appropriate parameter. Since the parameters will probably vary for VHF, HF, and satellite work, entire station setups could be saved on disk and then recalled when needed rather than having to change the parameters each time.

By having an option in the software to select different TNC interfaces, the same program could be used with TNCs with different command sets in much the same way as a word processor can be configured to print with different printers. Thus, a user would not have to learn a different command set for every TNC he might use.

Protocols

Protocols have been an area of concern and development since the beginning of packet radio. Today, the physical and data link layers are well defined and established as a result of many years of work.

Now it's time for the next step up the ladder with the network, transport, session, and possibly the presentation and application layers in the years ahead. There are many many protocols in existence which might be acceptable to the amateur packet radio environment. The problem is knowing what features we need, what features we do not, and the best way to implement them. The OSI/RM provides an excellent frame of reference for designing a network, but it is not cast in stone. Therefore, it is possible that in the future, amateur packet radio might handle the high level protocols in a different manner. But for now, we appear to be proceeding according to the OSI/RM.

Currently, most of the developmental work is being directed towards a network layer protocol. The network layer is discussed in detail in Chapter 4. Level 4 and 5 protocols are also under study, and there are several proposals, but no firm commitment at this time. Another area of development is the hardware necessary to implement these protocols such as the TAPR Network Node Controller.

NETWORKING

The future of packet radio will see many changes in the way we can use the amateur packet radio network. Packet operating practices will have to change to reflect the many new capabilities the amateur packet radio system will gain over time. Currently, our operating practices for the existing packet network are causing severe congestion in heavily populated areas. This is due, in most part, to our current practice of overusing one or two channels.

Because most BBSs are on the same channel (for mail forwarding), most users tend to congregate on the same channel out of habit or to monitor BBS activity. Digipeaters are on the frequency to allow for expanded mail forwarding and increased communications range. The current tendency is to have much of the activity taking place on the BBS channel. When new users arrive, there is no other active channel so they join in. As a result, the congestion is unbearable and at times almost no traffic can get through.

The solution to this problem is obvious, but due to a number of reasons is very difficult to enact. By spreading out the activity over a number of channels, the congestion on each channel would be much lower. In one approach, 2-meter FM channels could be left to individual users for keyboard-to-keyboard communications, file transfers, and experimentation. A less used frequency such as 220 MHz or 440 MHz could be chosen for BBS forwarding.

With this system, using current technology, future congestion problems can be reduced by expanding the number of channels proportional to the number of users. One disadvantage is that a single station cannot monitor activity on all the channels at once as was done when all activity occurred on a single channel. But that is a small price to pay for the reduced congestion.

The network will need to adapt as new hardware is implemented (such as high-speed modems). High-speed modems could be used first for BBS forwarding and long haul linking. In the transition period, these foreign modems would thus be hidden from the individual users.

One potential problem with a network based on many users on different channels in the same area is linking with other stations using current digipeater technology. Each user will have to keep a list of what digipeaters on what channel will take him or her to the desired destination. And if there is a choice of two or more paths, how do you know which is the best to choose? And how do you know which channel your friends are on at any given time?

These problems may be solved by the addition of level 3 packet switches (network nodes). The use of network nodes will give rise to the local area packet network. Users in a certain area will operate on the same frequency. Other users in a separate area will operate on a different frequency so they won't interfere. Users in one area will communicate with users in another area through the network nodes. It will not matter if the area is a few miles away, a few hundred miles away, or a few thousand miles away; the network node will choose the optimum route to get the data to the proper destination. The route could include VHF, UHF, satellite, HF, or a combination thereof.

All the individual user will need to know is the location of the station he or she wishes to communicate with; the network node will determine how to reach it. VHF DX operation will no longer be necessary, nor will connecting to a station through an inordinate number of digipeaters. Digipeaters may still be necessary for users who cannot reach their node directly.

The users in a given area will form a *Local Area Network (LAN)* with their own BBS and other facilities. Users will know where to reach their friends. With HF and satellites included in the network, worldwide networking will be possible.

The technology certainly exists to make this system a reality. All that is needed are funds, time, and volunteers. This network will make amateur packet radio a reliable and efficient means of worldwide communications.

PACKET CONTESTING

Packet will probably never become a contesting mode. The links on VHF and UHF should be so reliable that there is no room for a contest. Although on HF, there are many DX stations and it is possible to work Worked All States (WAS) and Worked All Countries (WAC) the potential for extensive operator abuse exists. Packet is too easy to automate. A program could be written to monitor all the callsigns received and an automatic QSO conducted with those that are of interest. If HF packet were to become a contesting mode, it most probably would have many restrictions. Most likely, QSOs would have to be live keyboard-to-keyboard and have a direct connection without digipeaters.

The purpose of contesting is to test how effective the contestors can operate their station, but a great deal has to do with how powerful their stations are, how big their antennas are, and how good the band conditions are. Most of these items have little relevance in amateur packet radio. Packet radio should be reliable under any band conditions and small, low-power stations ought to have as much access to the network as larger, more powerful ones. In fact, there is really no reason for the individual user to have a high-power packet station if he or she has reliable access to the network. So, packet radio is just not meant to be a contesting mode in the normal sense.

There are challenges that are well-suited to packet radio though. These would be one-time capability oriented challenges. A good example of this is the Golden Packet challenge. The first group to establish a terrestrial VHF link including digipeaters from the East Coast to the West Coast wins the contest. Of course, once the contest is won, it is over for good. These ability contests are beneficial to packet radio and provide challenges to spur future development.

CONCLUSION

Packet radio is a rapidly changing mode. It is no longer a revolution with few participants but undergoing an evolution with thousands of avid users. Each step that is made in developing packet radio helps to achieve the ultimate goal of a worldwide, error-free, efficient amateur radio data communications system. Each user adds to packet and helps to shape its future. Most progress in packet radio has been due to dedicated work by individuals and groups of individuals and will continue to be in the future.

This chapter is an appropriate ending to this book. I hope you recognize the potential of packet radio and have formulated your own ideas of what you would like to get out of packet and what you would like to contribute to it. Understanding the theory behind packet radio is meaningless unless you do something with it. I wish you the best of luck with amateur packet radio and hope that it is as much fun for you as it has been for me.

Appendix A

ASCII Chart

The chart on the next page provides the decimal and hexadecimal American Standard Code for Information Interchange (ASCII) codes from 0 to 127. It may prove useful when setting the control characters on a terminal or TNC.

ASCII Chart

Table A-1. ASCII Codes From 0 to 127.

DEC	HEX	ASCII	DEC	HEX	ASCII	DEC	HEX	ASCII
0	00	NUL	42	2A	*	85	55	U
1	01	SOH(A)	43	2B	+	86	56	V
2	02	STX(B)	44	2C	'	87	57	W
3	03	ETX(C)	45	2D	—	88	58	X
4	04	EOT(D)	46	2E	.	89	59	Y
5	05	ENQ(E)	47	2F	/	90	5A	Z
6	06	ACK(F)	48	30	0	91	5B	[
7	07	BEL(G)	49	31	1	92	5C	\
8	08	BS (H)	50	32	2	93	5D	]
9	09	HT (I)	51	33	3	94	5E	^
10	0A	LF (J)	52	34	4	95	5F	—
11	0B	VT (K)	53	35	5	96	60	'
12	0C	FF (L)	54	36	6	97	61	a
13	0D	CR (M)	55	37	7	98	62	b
14	0E	SO (N)	56	38	8	99	63	c
15	0F	SI (O)	57	39	9	100	64	d
16	10	DLE(P)	58	3A	:	101	65	e
17	11	DC1(Q)	59	3B	;	102	66	f
18	12	DC2(R)	60	3C	<	103	67	g
19	13	DC3(S)	61	3D	=	104	68	h
20	14	DC4(T)	62	3E	>	105	69	i
21	15	NAK(U)	63	3F	?	106	6A	j
22	16	SYN(V)	64	40	@	107	6B	k
23	17	ETB(W)	65	41	A	108	6C	l
24	18	CAN(X)	66	42	B	109	6D	m
25	19	EM (Y)	67	43	C	110	6E	n
26	1A	SUB(Z)	68	44	D	111	6F	o
27	1B	ESCAPE	69	45	E	112	70	p
28	1C	FS	70	46	F	113	71	q
29	1D	GS	71	47	G	114	72	r
30	1E	RS	72	48	H	115	73	s
31	1F	US	73	49	I	116	74	t
32	20		74	4A	J	117	75	u
33	21	!	75	4B	K	118	76	v
34	22	"	76	4C	L	119	77	w
35	23	#	77	4D	M	120	78	x
36	24	$	78	4E	N	121	79	y
37	25	%	79	4F	O	122	7A	z
38	26	&	80	50	P	123	7B	{
39	27	'	81	51	Q	124	7C	|
40	28	(	82	52	R	125	7D	}
41	29	)	83	53	S	126	7E	~
			84	54	T	127	7F	DELETE

Appendix B
The RS-232 C and D Standards

The RS-232 communications standard, originally developed in the late 1960s, remains the standard interface specification for serial asynchronous communications equipment. Other standards have been developed that have advantages over RS-232, but the new standards have failed to supplant RS-232 because of its widespread use. The reason we, as amateur radio operators, are concerned with the RS-232 standard is because most microcomputers and terminals, as well as TNCs, multimode units, and other modem devices that are used in amateur digital communications, support RS-232 communications.

In serial asynchronous communications, digital signals are sent as groups of a specified length sequentially on a single channel, and uneven intervals between transmissions are allowed. The digital signals usually represent characters in text. There are several standardized codes in use today for the transmission of text data; the American Standard Code for Information Interchange (ASCII) is the most common.

At first the RS-232 standard might seem a blessing because it seems to solve our serial communication needs very easily. After all, if two units each support RS-232, we can just wire them together and they will work; no problem. However, it is not that easy. Some so-called RS-232 compatible devices are not really very compatible. Compatibility is a matter of degrees; some devices are more compatible than others. This is definitely the case with most RS-232 implementations.

Some devices only support a subset of the RS-232 standard. Others vary pin assignments or voltage levels to meet their own requirements. Because of these potential problems, this appendix examines what the "standards" really standardize.

You will probably encounter the RS-232 standard when connecting your hardware packet Terminal Node Controller (TNC) or other digital modem device to your terminal or computer. Should you run into problems with the interface, a thorough understanding of the RS-232 standard will help you to diagnose the problem and come up with a solution. A knowledge of the standard can also be very helpful when wiring RS-232 cables.

INTRODUCTION

The proper name for the RS-232 standard is "Interface Between Data Terminal Equipment and Data Communications Equipment Employing Serial Binary Data Interchange." The standard was developed by the Electronic Industries Association (EIA) and the latest version is D; thus the reference to RS-232D. However, because version D of the RS-232 standard is relatively new, you will continue to see references to version C for the foreseeable future. Comité Consultatif Internationale de Télégraphique et Téléphonie (CCITT) Recommendation V.24 is almost identical to the RS-232C standard. This appendix will concentrate on the RS-232C standard because that version is currently being referenced by most microcomputer and peripheral manufacturers.

The RS-232C standard covers four main areas: the mechanical characteristics of the interface; the electrical signals across the interface; the function of each signal; the subsets of signals used for certain applications.

Data Terminal Equipment (DTE) and Data Communications Equipment (DCE), sometimes referred to as Data Circuit-terminating Equipment, are the two device classifications in RS-232. A DTE is a terminal, a computer, or any device capable of transmitting and receiving data. A DCE is a device that establishes, maintains, and terminates a connection. A DCE also provides any necessary signal conversion between the data it receives from and sends to the terminal and the data it sends and receives over the communications channel. Telephone modems and packet radio TNCs are DCE devices. We will learn more about the physical differences between DTEs and DCEs later in this section.

There is no specific connector in the RS-232C standard. However, the DB-25 connector is most commonly used and is now included in the RS-232D standard. Virtually all hardware TNCs use the DB-25 connector as the port for terminal communications. The DCE usually has the female connector, DB-25S (Socket). The maximum recommended cable length is 50 feet, and the maximum cable capacitance is 2500pF. Cable runs of greater than 50 feet are appropriate provided the load capacitance measured at the interface point and including the signal terminator does not exceed 2500pF.

SIGNALS

RS-232 electrical signals and their functions are referred to by four different systems: pin number, EIA designation, CCITT designation, and abbreviation of signal description. Following is information about the electrical signals most encountered when interfacing a DTE terminal to a DCE device. Fortunately, the full set of RS-232 signals is rarely used; so we are able to overlook numerous signals without worry.

Pin 1 is referenced by the EIA as AA, the CCITT as 101, and the abbreviation GND. It serves as the chassis ground between the two devices. However, it should not be depended upon for shock protection. But, this pin should definitely be connected at each end because opening in the chassis ground can cause problems that are very difficult to trace.

Pin 7 is referenced by the EIA as AB, the CCITT as 102, and the abbreviation SG. It serves as the signal ground. Pin 7 is the reference for all other pins and completes the circuit for the flow of current.

Pin 1 and pin 7 are the only two ground pins. Both should be connected; however, in most devices, pin 1 and pin 7 are connected to the same ground in the equipment. Thus, it is usually possible to get by with only one of the two connected. If there are separate chassis and signal grounds, and pin 1 is not wired and the ground at each device is at different potentials, current may flow through pin 7 and possibly interfere with data flow.

In version D of the RS-232 standard, pin 1 is defined as Shield and should not be connected to the interface. In version D, pin 1 is used to permit shielding of the interface cable. Pin 7, the signal ground, is the only ground connection you should make in the interface cable when using version D.

Pin 2 is referenced by the EIA as BA, the CCITT as 103, and the abbreviation TD. This pin serves as the Transmit Data pin. All information sent via the RS-232 port comes out on this pin.

Pin 3 is referenced by the EIA as BB, the CCITT as 104, and the abbreviation RD. This pin serves as the Receive Data pin. All data received via the RS-232 port comes in on this pin.

These descriptions are viewed from the DTE. DCE sends data on pin 3 and receives on pin 2. Thus, the DTE transmits on pin 2 and the DCE receives on pin 2, and the DCE transmits on pin 3 and the DTE receives on pin 3.

Pin 4 is referenced by the EIA as CA, the CCITT as 105, and the abbreviation RTS. This pin serves as the Request To Send indicator. When the DTE has data to send, it asserts the RTS.

Pin 5 is referenced by the EIA as CB, the CCITT as 106, and the abbreviation CTS. This pin serves as the Clear To Send indicator. The DCE assets the CTS when it is able to receive data from the DTE. According to the standard, the CTS may only be asserted after receiving a RTS from the DTE.

Pin 6 is referenced by the EIA as CC, the CCITT as 107, and the abbreviation DSR. This pin serves as the Data Set Ready indicator. It is asserted as a response to the DTR signal and indicates that the DCE is ready for operation.

Pin 20 is referenced by the EIA as CD, the CCITT as 108/2, and the abbreviation DTR. This pin serves as the Data Terminal Ready indicator. The DTR indicates that the DTE is ready to send and receive data. The DTR is asserted whenever the DTE has data to send, or in some cases whenever the terminal is operating.

Pin 8 is referenced by the EIA as CF, the CCITT as 109, and the abbreviation DCD. It serves as the Data Carrier Detect (or just Carrier Detect) indicator. The DCE asserts this pin when the communications channel is ready. Many DTE will not transmit or receive data unless this pin is asserted. In some cases, pin 8 is wired to pin 20 so it is always asserted.

When the RS-232C standard is properly implemented, data will not be transmitted unless the RTS, CTS, DSR, DTR, and DCD pins are asserted. There are many other RS-232C signals, 20 in all, but the ones listed here are the most commonly used.

SIGNAL LEVELS

RS-232 signal voltages are not compatible with those used by most computer circuitry, so an additional power supply is incorporated in RS-232 equipment to provide the necessary voltages. RS-232 signals are referenced to the pin 7 signal ground. The positive voltages can range from 5 to 25 volts. On pins 2 and 3, a positive voltage indicates a logic 0 level. The negative voltages range from -5 to -25 volts. On pins 2 and 3, a negative voltage indicates a logic 1 level. The polarities are reversed for the control line logic levels with a logic level 1, meaning the pin is asserted on.

When transmitting, voltages of $+12$ and -12 volts are usually used by most devices. When receiving, the positive voltage must be greater than 3 volts and the negative voltage must be less than -3 volts in order to be correctly interpreted by the receiving circuitry.

CABLE CONFIGURATIONS

The following section describes several examples of RS-232 cables for a variety of applications. They might be of help when you are wiring your own cables or attempting to diagnose problems with your cable and interface.

Minimum Cable

In a minimum RS-232 cable, as few as three pins might be connected. This is very convenient for use in lengthy cable runs. Pins 1(GND), 2(TD), 3(RD), and 4(SG) are connected. (See Fig. B-1.) If the signal and chassis ground are connected in the equipment, only one pin is necessary; pin 1 is usually chosen. In order for this cable to work, the RTS/CTS and DSR/DTR pairs must be ignored.

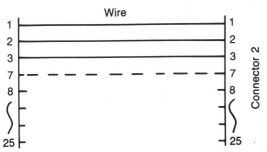

Fig. B-1. Diagram of a RS-232 minimum cable.

182 The RS-232 C and D Standards

If a piece of equipment will not work unless a pin is asserted, just wire it directly to the voltage required. Software flow control must be used with this three-wire cable; more on flow control later.

Full Cable

This cable should provide all connections necessary for most RS-232C applications. The following pins should be connected: Pin 1 or pin 7, pin 2, pin 3, pin 4, pin 5, pin 6, pin 8, and pin 20. (See Fig. B-2.) This cable works in most situations and allows for hardware flow control.

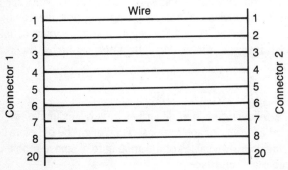

Fig. B-2. Diagram of a RS-232 full cable.

NULL MODEM

A null modem adapter allows two DTEs or DCEs to communicate with each other. Several pins must be reversed to allow the proper signals to reach the proper pins. (See Fig. B-3.) If you recall from the signal descriptions of pins 2 and 3, a DTE transmits on pin 2 and receives on pin 3, and a DCE transmits on pin 3 and receives on pin 2. So when you are connecting two devices of a like type, these pins must be cross-connected so that one device's pin 2 is wired to the other device's pin 3 and the first device's pin 3 is wired to the other's pin 2. This allows the two devices to transmit and receive on their proper pins.

The RTS of each device should be wired to its own CTS and to the DCD of the other device. This allows a Request To Send to receive an instant Clear To Send and also asserts the Data Carrier Detect so the other device knows a transmission is coming. Additionally, the DTR and DSR pins must be cross-connected in the

Null Modem 183

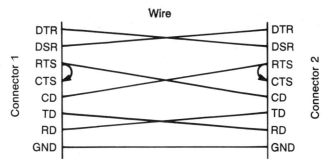

Fig. B-3. Diagram of a RS-232 null modem.

same manner as the TD and RD pins to allow for proper signaling. The pin 1 and 7 grounds are wired straight through as usual.

The null modem adapter is useful when transferring files between two computers or when connecting a computer to a printer that is wired as a DTE. It can be incorporated into a cable by switching the cable wires or through an adapter that is inserted into the cable. (See Fig. B-4.)

Fig. B-4. A Null modem adapter.

FLOW CONTROL

Flow control is one of the potential problem areas of the RS-232 interface. *Flow control* is the process of stopping and starting the flow of data between devices. Flow control can be implemented in hardware, using the interface's signals, or in software. Under software flow control, the data flow is controlled independently of the physical interface.

The RS-232 standard was not designed for hardware flow control. Rather, it expected all flow control to be provided through software. One method of software flow control, known as Xon/Xoff, works by sending control characters over the physical interface. However, many computer manufacturers have attempted to control the flow of data between the devices through use of the RTS/CTS and DTR/DSR pairs. This has been accomplished with varying degrees of success with printers and modems, and has been carried over to packet radio TNCs and other amateur digital communications interfaces.

According to the RS-232 standard, the DCE is not allowed to drop the CTS until the DTE drops the RTS. The DCE should not drop its CTS at random for flow control. The proper use of the RTS/CTS pair is to allow the terminal to request use of the communication channel.

However, hardware flow control using these pins is possible. Many TNCs don't have full RTS/CTS capability, so they can be used for flow control if the terminal will allow it, and most terminals will. Another approach, DTR/DSR flow control, works by turning off the DSR pin when the device can accept no more data.

There is no guarantee that hardware flow control will work, but it usually does. Many problems can occur when using hardware flow control, such as when to stop accepting data, what to do with a character that is partially sent when the device is told to stop sending, and when and at what character to resume sending. Check the terminal and TNC manuals carefully if you are planning to implement hardware flow control. Generally, it is usually best to stick with Xon/Xoff software flow control.

NONSTANDARD IMPLEMENTATIONS

Another problem that can pop up when working with RS-232 "compatible" devices is nonstandard voltage levels. Instead of using the standard +12 and −12 volts, some equipment use voltages such as +5 and −5 or +5 and 0. Depending on how sensitive the RS-232 devices are to voltage levels, these nonstandard voltage levels may or may not work. In most cases they usually do.

Nonstandard pin assignments are yet another problem that can be encountered. Some TNCs have other uses assigned to some pins other than as specified by the standard. These pins are usually not used for input/output (I/O) and are usually ones that are not commonly used by most devices. These pins might have strange voltages on them or be used as special control lines. It is a good idea not to connect these pins unless you are certain they will not interfere with the data transmission or damage any equipment. Check the manuals for more specific information about a a particular pin's function.

Limitations

Although the RS-232 standard has been a great aid over the years in standardizing serial data communication between devices, it does have many limitations. These limitations are usually not of concern to us as amateur radio operators, but they should be considered when working with RS-232 communications.

The maximum recommended cable length of 50 feet is not often found to be very limiting for amateur digital communications applications. The value of 50 feet is derived by dividing the maximum capacitance of 2500pF by the capacitance of a foot of cable, which is usually about 50pF. The cable length can be increased through the use of shielded cable and in-line amplifiers.

The fact that the voltages used by RS-232 are not the same as those used to power most computer components requires the addition of another power supply.

RS-232 utilizes what is known as an *unbalanced ground*. There is only one signal ground for all the pins, and a difference in ground potential at each end can change the allowable voltage detection range. As a result, signal detection errors may occur.

Despite these drawbacks, the RS-232 standard is fine for limited-distance, medium-speed applications such as microcomputer communications.

WIRING CABLES

Wiring your own RS-232 cables can save you a great deal of money over the price of a completed cable, and you can modify the cable to meet your own needs very easily. There are three main decisions to be made when wiring RS-232 cables: connector type, cable type, and number and type of pins.

The most common RS-232 connector is the DB-25. The DB-25 connector comes in several varieties. (See Fig. B-5.) You must first determine whether your equipment needs male or female connectors. The male connector is known as DB-25P (Plug), and the female connector is known as DB-25S (Socket). Each plug comes in two varieties: solder type and friction type.

186   The RS-232 C and D Standards

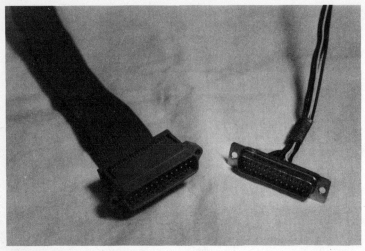

Fig. B-5. Solder and friction type DB-25 connectors.

With the solder type, individual wires are soldered to each needed pin. The cable can be composed of individual wires, usually a four-five-wire ribbon cable, or four-wire telephone cable. Hoods are protective covers designed to fit over the back of the solder type DB-25 connector and provide a convenient handhold for plugging and unplugging the connector.

The friction type, also called *insulation displacement*, connectors are for use with ribbon cable; usually 25-wire. The cable is simply placed in a slot on the back of the connector, and the connector is compressed. Connector pins puncture the ribbon cable's insulation and make contact with the wires. Friction connectors provide a convenient means to quickly wire all 25 pins.

When you are constructing a cable, it is a good idea to always add an extra foot or so of cable. In this way, you can rearrange the station much easier because you have not tied down the TNC to one particular location.

CONCLUSION

The RS-232C and RS-232D standards can be a pain to work with and diagnose if not working properly. In most cases, however, the interface works fine the first time. If you already have a RS-232 connection to a modem or other device, simply substitute the TNC (or whatever device you are using) for that device and everything should work fine.

Should you have problems with a RS-232 interface, it is important to determine whether the interface cabling, the TNC, or the terminal's hardware and/or communications software is at fault. If the fault is isolated to the interface cabling, the information in this appendix should make finding and solving the problem a little easier.

Appendix C

Sources

This appendix contains the addresses of most of the organizations, companies, and publications mentioned in this book. This appendix is segmented into several different areas (packet organizations, standards organizations, publications, and companies) to simplify locating a particular subject.

PACKET ORGANIZATIONS

Many of the organizations listed below publish newsletters and promote packet activity in their area.

AMRAD
Amateur Radio Research and Development Corp.
P.O. Drawer 6148
McLean, VA 22106

AMSAT
The Amateur Radio Satellite Corp.
850 Sligo #601
Silver Springs, MD 20910
(301) 589-6062

ARRL
American Radio Relay League
225 Main Street
Newington, CT 06111

CAPRA
Chicago Amateur Packet Radio Association
P.O. Box 8251
Rolling Meadows, IL 60008

EPROM
Eastern Packet Radio of Michigan
307 Ross Drive
Monroe, MI 48161

FADCA
Florida Amateur Digital Communications Association
812 Childers Loop
Brandon, FL 33511
(813) 689-3355

GRAPES
Georgia Radio Amateur Packet Enthusiast Society
P.O. Box 1354
Conyers, GA 30207

LAPG
Los Angeles Area Packet Group
P.O. Box 6026
Mission Hills, CA 91345

MAPR
Minnesota Amateur Packet Radio
565 Redwood Lane
New Brighton, MN 55112

MAPRC
Mid-Atlantic Packet Radio Counsel
6388 Guilford Road
Clarksville, MD 21029

NEPRA
New England Packet Radio Association
P.O. Box 15
Bedford, MA 01730

PPRS
Pacific Packet Radio Society
P.O. Box 51562
Palo Alto, CA 94303

RATS
Radio Amateur Telecommunications Society
North
206 North Vivyen Street
Bergenfield, NJ 07621
South
RD 2 Burnt House Road
Indian Mills, NJ 08088

RMPRA
Rocky Mountain Packet Radio Association
5222 Borrego Drive
Colorado Springs, CO 80901

SDPG
San Diego Packet Group
10230 Mayor Circle
San Diego, CA 92126

SLAPR
St. Louis Area Packet Radio
9926 Lewis and Clark
St. Louis, MO 63136

TAPR
Tucson Amateur Packet Radio
P.O. Box 22888
Tucson, AZ 85734
(602) 746-1166

TPRS
Texas Packet Radio Society
P.O. Box 831566
Richardson, TX 75083

VADCG
Vancouver Amateur Digital Communications Group
9531 Odlin Rd
Richmond, BC V6X1E1
Canada

STANDARDS ORGANIZATIONS

Write to the following organizations for information on specific standards.

CCITT (Comité Consultatif Internationale de Téllégraphique et Téllé'phonie)
International Telecommunications Union
General Secretariat
Sales Service
Place de Nation
CH 1211
Geneva 20
Switzerland

or

United Nations Bookstore
Room 32B
UN General Assembly Building
New York, NY 10017

EIA
EIA Engineering Department
Standard Sales
2001 Eye Street, N.W.
Washington, D.C. 20006

ISO
ISO Central Secretariat
Case Postale 56
1211 Geneva 20
Switzerland

PUBLICATIONS

The following publications are good for packet radio information.

192 Sources

AMRAD Newsletter
5829 Parakeet Drive
Burke, VA 22015

CQ Magazine
76 North Broadway
Hicksville, NY 11801

Gateway
ARRL
225 Main Street
Newington, CT 06111

Ham Radio Magazine
Greenville, NH 03048

The Packet
VADCG
9531 Odlin Road
Richmond, BC V6X1E1
Canada

Packet Status Register
TAPR
P.O. Box 22888
Tucson, AZ 85734

QST
ARRL
225 Main Street
Newington, CT 06111

Spec-Com
P.O. Box H
Lowden, IA 52255

73 Magazine
WGE Center
Peterborough, NH 03458

COMPANIES
The following companies were mentioned in this book.

AEA
Advanced Electronic Applications
2006 196th SW
Lynnwood, WA 98036

Bill Ashby and Son
P.O. Box 332
Pluckemin, NJ 07978

DRSI
Digital Radio Systems, Inc.
2065 Range Road
Clearwater, FL 34625

GLB Electronics
151 Commerce Parkway
Buffalo, NY 14224

HAPN
Box 4466, Station D
Hamilton, Ontario
Canada L8V 457

Heathkit
Amateur Radio Department
Benton Harbor, MI 49022

Kantronics
1202 East 23rd Street
Lawrence, KS 66044

MFJ Enterprises
921 Louisville Road
Starkville, MS 39759

Microlog
18713 Mooney Drive
Gaithersburg, MD 20879

Pac-Comm Packet Radio Systems
3652 W. Cypress Street
Tampa, FL 33607

Richcraft Engineering
1 Wahmeda Industrial Park
Drawer 1065
Chautauqua, NY 14722

S. Fine Software
P.O. Box 6037
State College, PA 16801

Software 2000
1127 Hetrick Avenue
Arroyo Grande, CA 93420

TAPR
P.O. Box 22888
Tucson, AZ 85734

Appendix D

Operating Frequencies

Amateur packet operation occurs on a variety of frequencies in the United States. The most common frequencies are listed in this appendix. See Chapters 3 and 6 for more information on transceiver specifications and operating information.

HF

HF operation usually conforms to the Bell 103 standard. However, the frequency of the tone pairs sometimes varies between units. Therefore, it may be necessary to tune slightly above or below the given frequencies. The frequencies given are the LSB carrier frequency with TAPR tone pairs (1600 Hz and 1800 Hz). HF operation takes place on a variety of frequencies with 14.103 MHz being the most popular. Table D-1 shows a list of HF packet frequencies.

VHF

VHF operation usually conforms to the Bell 202 standard. The vast majority of VHF packet activity takes place on 145.01 MHz. VHF packet operation occurs on unusual frequencies in some areas of the country for a variety of reasons. Table D-2 shows a list of VHF packet frequencies.

Operating Frequencies

Table D-1

Band	Frequency
80-meters	3.607 MHz
40-meters	7.093 MHz
30-meters	10.147 MHz
20-meters	14.103 MHz
	14.105 MHz
	14.107 MHz
	14.109 MHz
10-meters	28.102 MHz

Note: HF packet operation also occurs on standard RTTY frequencies.

Table D-2

Band	Frequency	Notes
2-meters	145.01 MHz	Nationwide
	145.03 MHz	Nationwide
	145.05 MHz	Nationwide
	145.07 MHz	Nationwide
	145.09 MHz	Nationwide
	145.36 MHz	Southern CA
	146.745/.145 MHz	
	147.555 MHz	Southern and central IL, IO, and east MS.
	145.70 MHz	Denver, CO
	145.57 MHz	Dallas, TX
	146.13/.73 MHz	Tucson, AZ
	146.13/.73 MHz	Atlanta, GA
1.25-meters	223.40 MHz	Nationwide
	223.42 MHz	Local
	223.44 MHz	Local
	223.46 MHz	Local
	223.48 MHz	Local

Appendix E

An Introduction to Amateur Radio

This appendix is included for those readers who are not familiar with the amateur radio service or who are not sure what license class is necessary to operate packet. Amateur packet radio is but a small part of the world of amateur radio. Amateur radio is a worldwide hobby which provides many services to the individuals and countries that encourage its growth. There are over one million licensed amateur radio operators, including over 400,000 in the United States.

INTRODUCTION

Amateur radio can be defined as radio communications between licensed stations without financial remuneration. Amateur radio operators have a wide variety of communications modes available for their use. The major modes in use today include CW (Continuous Wave or Morse code), SSB (Single Side Band—voice), FM (Frequency Modulation—voice), Television (both fast and slow scan), RTTY (radio teletype), and, of course, packet. Communications can be by line-of-sight, ionosphere reflection, satellite, or more exotic modes such as moonbounce and meteor scatter.

Radio amateurs worldwide are authorized to operate on certain frequencies just as commercial and government stations are. The frequency space occupied by amateur radio is very valuable, and the use of this space would be welcomed by commercial users and governments. Amateur radio justifies its utilization of these frequencies by providing emergency communications and contributing to the development of new communications techniques. Packet radio fits right in here.

In the United States, a license issued by the FCC (Federal Communications Commission) is required to operate an amateur radio station. There are several grades of licenses available. Each increase in license grade provides additional capabilities and requires the user to pass more difficult tests to obtain the license. These tests assure that the operators are proficient enough to operate a station in the modes allowed.

LICENSES

Licenses, capabilities, and testing vary from country to country. Five amateur radio licenses are available in the United States; here's a brief summary of the elements tested and the capabilities.

The Novice license is the easiest license to obtain. Its license privileges were recently expanded to include voice and digital communications privileges on one HF band and several VHF and UHF bands in addition to the previously permitted CW operation on the HF bands. The testing is relatively simple: Morse code at 5 wpm (words per minute), operating techniques, and basic electronic and radio theory.

The Technician class license allows all the capabilities of a Novice plus additional VHF and UHF capabilities. The test consists of the Novice elements plus a higher-level theory test.

The General class license allows all the capabilities of a Technician plus increased operating capabilities on certain HF frequencies. The testing is the same as the Technician with the exception of a 13 wpm Morse code test. This is the most popular license in the U.S.

The Advanced class license allows all the capabilities of a General plus additional HF frequency privileges. The testing is the same as the General with the exception of a higher level theory test.

The Extra class license is the highest class of license available. It offers all the capabilities of the Advanced class plus the highest level of HF frequency operating privileges. The testing includes all the theory elements of the Advanced class plus another high-level theory test and a 20 wpm code test.

Help Us Help You

So that we can better provide you with the practical information you need, please take a moment to complete and return this card.

1. I am interested in books on the following subjects:
- ☐ architecture & design
- ☐ automotive
- ☐ aviation
- ☐ business & finance
- ☐ computer, mini & mainframe
- ☐ computer, micros
- ☐ other_____
- ☐ electronics
- ☐ engineering
- ☐ hobbies & crafts
- ☐ how-to, do-it-yourself
- ☐ military history
- ☐ nautical

2. I own/use a computer:
- ☐ Apple/Macintosh_____
- ☐ Commodore_____
- ☐ IBM_____
- ☐ Other_____

3. This card came from TAB book (no. or title):
_____

4. I purchase books from/by:
- ☐ general bookstores
- ☐ technical bookstores
- ☐ college bookstores
- ☐ mail
- ☐ telephone
- ☐ electronic mail
- ☐ hobby stores
- ☐ art materials stores

Comments _____

Name _____

Address _____

City _____

State/Zip _____

TAB BOOKS Inc.

BUSINESS REPLY MAIL
FIRST CLASS   PERMIT NO. 9   BLUE RIDGE SUMMIT, PA 17214

POSTAGE WILL BE PAID BY ADDRESSEE

TAB BOOKS Inc.
Blue Ridge Summit, PA 17214-9988

NO POSTAGE
NECESSARY
IF MAILED
IN THE
UNITED STATES

CALLSIGNS

Every licensed operator is assigned a unique callsign. New callsigns in the U.S. have a different format for each license class. In the case of an upgrade, the amateur may opt to keep his or her previous callsign from a lower level license. Currently issued Novice callsigns are in the format of XX#XXX, Technician and General are X#XXX, Advanced is XX#XX, and Extra is X#XX or XX#X (where X indicates a letter and # a number). All U.S. callsigns can be identified by the prefix (the portion of the call that procedes the number). U.S. callsigns always have a N, K, A, or W for the first letter.

The number in the callsign represents the district in which the license was issued. The U.S. is divided into numbered districts (0 to 9). When a licensee moves from one district to another, he or she may keep his or her present call or apply for a new one with the new district number. The new call will not contain the same alphabetical characters as the old call since new calls are issued in alphabetical order.

PACKET OPERATION

The minimum license class currently necessary to operate amateur packet radio in the United States is a Novice class license. However, Novices are not permitted any privileges on the 2-meter band where most VHF amateur packet radio activity takes place. A Technician class licensee may operate packet on 2-meters; however, both Technician and Novice licensees may not operate HF packet directly (with the exception of the 10-meter band). A Novice or Technician may operate through a packet gateway station onto the HF bands from the allowed VHF frequencies provided the gateway is under the control of an appropriate class licensee for the HF frequency being used. General, Advanced, and Extra class licensees may operate packet on all bands.

CONCLUSION

Amateur radio is a diverse hobby. I have not attempted to fully describe it in the limited space available here. For more information, contact the ARRL; their address is listed in Appendix C. Also, look for an amateur radio magazine on the newsstand. Check to see if there are any amateur radio dealers nearby; if so, drop by for a visit.

Glossary

This glossary is designed to provide you with brief definitions of a comprehensive list of packet radio related terms. Check the index to see if the term you are looking up is contained in one of the chapters.

ACK Abbreviation for acknowledgement. Sent by the destination station back to the originating station to indicate the successful reception of a frame. (*See* NAK.)

address 1. The specific designation given to each station on the network for identification purposes. In AX.25, the address consists of the amateur station's callsign plus a substation identifier ranging from 0 to 15. 2. The first field of an HDLC frame following the initial flag which contains the addresses of the originating and destination station. In AX.25, the address field may include digipeaters. (*See* SSID.)

AEA Acronym for Advanced Electronics Applications, Inc. A commercial manufacturer of amateur packet radio products.

AFSK Acronym for Audio Frequency Shift Keying. A method of modulation in which the RF carrier frequency remains constant and an audio modulation tone is shifted in frequency. When used on an SSB transmitter, it cannot be differentiated from FSK.

ALOHA Also **ALOHANET.** An early packet radio network set up at the University of Hawaii in 1970 for research and development of packet radio communications. (*See* menehune.)

alpha testing The first stage of product testing after the prototype development. The sequence is: conception, design, prototype, alpha testing, beta testing, and production.

AMRAD Amateur Radio Research And Development Corporation. A nonprofit organization involved in amateur packet radio development.

AMSAT The AMateur SATellite Corporation. A United States-based non-profit organization committed to the development and encouragement of amateur satellite activity.

AMTOR AMateur Teletype Over Radio. An advanced form of RTTY usually operated on the HF bands.

analog A signal that varies in a continuous manner (e.g., voice, music, and voltage and currents that vary in a continuous manner). (*See* digital.)

ANSI Acronym for American National Standards Institute. The principal standards development organization in the United States. (*See* CCITT, EIA, ISO.)

application layer Level 7 of the OSI/RM. Contains user software.

argument A variable expression that follows a command.

ARRL Acronym for American Radio Relay League.

ASCII Acronym for American Standard Code for Information Interchange. Also USASCII. A seven-bit code established by ANSI to achieve compatibility between digital devices.

assembly language A low-level, high-speed computer language consisting of mnemonics and operands which are converted, or assembled, to machine code. (*See* machine code, BASIC, C.)

asynchronous Also called Start-Stop transmission. Digital signals which are sent as groups of a specified length with start and stop bit indicators at the beginning and end of each group. Usually used when time intervals between transmitted groups may be uneven. (*See* synchronous.)

AX.25 Amateur packet radio protocol version of the X.25 protocol. Usually used in reference to the data link layer protocol in use by most amateur packet stations.

BASIC Acronym for Beginner's All-purpose Symbolic Instruction Code. A high-level computer language included with most microcomputers. (*See* C, assembly language, machine code.)

Glossary

baud A unit of signaling speed equal to the number of discrete signal events per second. Baud is the same as BPS only if each signal event represents exactly one bit.

Baudot Also called *Murray* code. A five-level code for the transmission of data in digital form. Named for Emile Baudot. Baudot code is usually found in older teleprinters.

BBS Also called CBBS, PBBS, Mailbox. Acronym for Bulletin Board System. An automated computer system which can be controlled from a remote location. Usually capable of sending and receiving messages and files.

Bell 103 A modem standard with a 200 Hz shift (1070 Hz, 1270 Hz) operating at 300 baud. Used for HF amateur packet radio operation.

Bell 202 A modem standard with a 1000 Hz shift (1200 Hz, 2200 Hz) operating at 1200 baud. Used for VHF FM amateur packet radio operation.

BER Acronym for Bit Error Rate.

beta testing The fifth and usually final stage of product development before general production and release. The sequence is: conception, design, prototype, alpha testing, beta testing, and production.

binary A number system based on the powers of 2. The only characters are a "0" and a "1." Binary digits are easily transmitted and stored in electronic equipment. (*See* bit, hex, octal.)

bipolar keying A technique in which a binary "1" is represented by a positive pulse and a binary "0" is represented by a negative pulse. Bipolar keying is the system used by NRZI on amateur packet radio. (*See* NRZ, Manchester, encoding technique.)

bit Abbreviation for BInary DigiT; either a "0" or a "1."

bit stuffing The addition of a binary "0" following all sequences of five binary "1"s in the data transmission to avoid the accidental occurrence of a flag. The added binary "0" is removed by the receiving station.

BOP Acronym for Bit Oriented Protocol. Control and data consume only as many bits as needed. No minimum binary grouping. (*See* COP.)

buffer Memory space set aside for the temporary storage of data until recalled, processed, or permanently stored.

bus network A network configuration in which all nodes are on the same channel and may communicate with each other directly providing they are within range.

byte A grouping of eight bits. (*See* nybble, octet.)

C A high-speed, high-level computer language. (*See* BASIC, machine code, assembly language.)

CCITT Comite Consultatif Internationale de Telegraphique et Telephonie. An international committee that establishes international communications standards. (*See* ANSI, EIA, ISO.)

checksum The numeric result of a CRC. Sent within a frame as the FCS.

collision When two or more stations transmit at the same time, or when one or more stations transmit while another station is transmitting. A collision may destroy one or more of the transmissions depending on the relative strength of the signals and the sensitivity of the receivers.

command A string of characters recognized and acted upon by a device.

command set A subset of the user interface consisting of all available commands and parameters. May be organized by group, function, or alphabetically.

connection The condition of having established communications between two stations via a preset path.

contention A condition in which two or more stations try to transmit at the same time.

control character A special character recognized by the receiver (usually a computer) as having a special meaning. Usually sent by pressing a control key and the appropriate character on the keyboard. Control characters are written in the abbreviated form as CTRL or CTL. For example, CTRL-C: break, CTRL-M: carriage return.

COP Acronym for Character Oriented Protocol. All information must be sent as characters of a specified length. COPs are usually less efficient and less flexible than BOPs. (*See* BOP.)

CPU Acronym for Central Processing Unit. The "brains" of a computer. Responsible for directing the flow of data throughout the computer.

CRC Acronym for Cyclic Redundancy Check. An error detection scheme in which a check character is generated by dividing the entire numeric binary value of a block of data by a generator polynomial expression. The CRC value is sent along with the data, and at the destination station, the CRC is recomputed from the received data. If the received CRC value matches the one generated from the received data, the data is considered error-free. (*See* FCS, checksum.)

CRRL Acronym for Canadian Radio Relay League.

CSMA/CD Acronym for Carrier Sense Multiple Access with Collision Detection. The system used in amateur packet radio to handle TDM of the channel and contention. Each station monitors the channel and only transmits when the channel is clear. The absence of an ACK from the destination station indicates that a collision may have occurred, and the transmission is resent after waiting a radom time interval.

CTS Acronym for Clear To Send in the RS-232 standard. Also referred to as Pin 5, CB by the EIA, and 106 by the CCITT.

data The digital information which is being transmitted or received.

data link layer Level 2 of the OSI/RM. Arranges bits into frames, establishes and maintains a link, and performs error detection and recovery. ISO HDLC is the most common level 2 protocol.

datagram A type of packet networking in which each packet contains complete and extensive addressing and control information. This allows for variable routing at the expense of greater overhead. (*See* virtual circuit.)

DB-25 A series of 25-pin connectors which are commonly used in RS-232 interfacing applications.

DCD 1. An indicator which signals the presence of a signal. 2. Acronym for Data Carrier Detect in the RS-232 standard. Also referred to as Pin 8, CF by the EIA, and 109 by the CCITT.

DCE Acronym for Data Communication Equipment. A device capable of establishing, maintaining, and terminating a connection. Also may have to handle signal conversion and coding. (*See* DTE.)

demodulation The process of retrieving data from a modulated signal. (*See* modulation, modem.)

digipeater A simplex packet repeater which stores an incoming packet, and, if so instructed, retransmits it. The digipeater does not retain a copy of the packet once sent or wait for an acknowledgement from the next node.

digital A discrete or discontinuous signal whose various states are identified with specified values. (*See* analog, RS-232, TTL.)

DOS Acronym for Disk Operating System. A program which handles all disk access by the computer system. May be loaded into memory or permanently stored in firmware.

downlink A radio link originating at a satellite and terminating at a ground station.

DSR Acronym for Data Set Ready in the RS-232 standard. Also referred to as Pin 6, CC by the EIA, and 107 by the CCITT.

DTE Acronym for Data Terminal Equipment. A device capable of I/O. (*See* DCE.)

DTR Acronym for Data Terminal Ready in the RS-232 standard. Also referred to as Pin 20, CD by the EIA, and 108/2 by the CCITT.

dumb terminal A communications terminal with only the basic capabilities necessary for communications such as an input device, an output device, and a predefined I/O port. (*See* smart terminal.)

EBCDIC Acronym for Extended Binary Coded Decimal Interchange Code. An eight-level character code developed by IBM and used primarily in their equipment. (*See* ASCII, Baudot.)

EIA Acronym for Electronic Industries Association. A standards organization specializing in interface equipment. (*See* ANSI, CCITT.)

encoding technique Also called line coding, channel coding, and data format. The system utilized to encode the digital data for transmission. (*See* NRZ, NRZI, Manchester.)

FAD Acronym for Frame Assembler/Disassembler. Often used interchangeably with PAD and TNC.

FADCA Acronym for Florida Amateur Digital Communications Association. Published the *FADCA > BEACON* newsletter, and *Packet Radio Magazine*.

FCS Acronym for Frame Check Sequence. A CRC for a frame.

FDM Acronym for Frequency Division Multiplexing. A technique for distributing users over a number of separate channels; each channel may have different characteristics. (*See* TDM, SDM.)

flag A unique binary sequence used to delimit frames at the data link layer. In HDLC, the flag is 01111110. (*See* bit stuffing.)

flow control The process of stopping and starting the flow of data between devices.

frame A group of bits delimited by flags. May contain control information and data.

FSK Acronym for Frequency Shift Keying. A method of frequency modulation in which the frequency varies. (*See* AFSK, PSK.)

full duplex Simultaneous two-way independent transmission in both directions on separate channels. (*See* simplex, half duplex.)

Gateway The amateur packet radio newsletter published biweekly by the ARRL.

gateway A device which retransmits received data in another format or on another channel.

half duplex A circuit designed for transmission in either direction on two separate channels but not both directions simultaneously. (*See* full duplex, simplex.)

hard drive Also called *Winchester drive*. A rigid disk magnetic storage device permanently sealed in a hermetic container which can store large amounts of data; usually (for microcomputers) 20 to 60 megabytes.

hardware Physical equipment as opposed to a program or protocol; for example, TNC board, computer, printer. (*See* software.)

HDLC Acronym for High-level Data Link Controller. An ISO standard for the data link layer of the OSI/RM. (*See* protocol.)

header The collective components of a frame preceding the information component. The header of an AX.25 frame consists of an opening flag, the address field, and control information.

HEX Abbreviation for hexadecimal. A number system based on powers of 16. Characters are 0-9 and A-F. (*See* binary, octal.)

IBM Acronym for International Business Machines Corporation.

IC Acronym for Integrated Circuit.

I/O Acronym for Input/Output. Used in reference to any system or function that deals with sending and receiving data.

ISO Acronym for International Standards Organization. (*See* CCITT, EIA, ANSI.)

JAMSAT The Japanese AMSAT affiliate.

JAS-1 A packet radio satellite designed by JAMSAT which features a flying mailbox and digipeater capabilities. (*See* RUDAK, Pacsat.)

kludge A temporary "quick and dirty" solution to a problem. Often used to imply that a system is inefficient, ill-designed, and in need of improvement.

landline Common expression for a terrestrial cable link between two stations. Usually used in reference to the telephone system.

LAPB Acronym for Link Access Procedure Balanced. A subset of HDLC in which each node is treated on an equivalent basis with each able to send both commands and responses.

LCD Acronym for Liquid Crystal Display. A display device commonly used in portable computers. Contains a crystalline liquid whose optical properties change in the presence of an electric field to appear either light or dark. Must have an external light source to be visible.

level 1-7 Numerical designators for the OSI/RM levels as follows; level 1: physical layer, level 2: datalink layer, level 3: network layer, level 4: transport layer, level 5: session layer, level 6: presentation layer, and level 7: application layer.

machine code A low-level, high-speed computer language consisting of the actual binary instructions acted upon by the computer. (*See* assembly language, BASIC, C.)

mailbox (*See* BBS.)

Manchester Two types are Manchester I and Manchester II. An encoding technique similar to NRZI, but differs because the transition from positive to negative or negative to positive occurs in the middle of the bit signaling period. (*See* NRZ.)

MAPRC Acronym for Mid-Atlantic Packet Radio Council. A packet group based in the mid-Atlantic area of the continental United States.

menehune The central node in the Aloha packet radio network.

Modem Contraction for MOdulator/DEModulator. A device that modulates transmitted signals and demodulates received signals. It serves as an interface between an analog communications system and digital devices.

modulation The process of adding a signal to a carrier to transmit information. Can be used in reference to voice communications, but refers to digital data in the context of packet radio. (*See* demodulation, modem.)

monitor mode A mode in which the TNC is instructed to forward all received packets to the terminal. The user usually specifies categories of packets to be received (i.e., to or from certain stations, digipeated packets, control packets.

MSK Acronym for Minimum Shift Keying. A modulation method similar to FSK in which the shift in hertz is equal to half the signaling rate in BPS. (*See* AFSK, FSK, PSK.)

multiconnect The ability of a packet station to connect with more than one station simultaneously.

multiplex The process of dividing a communications medium so that many users can share it. (*See* FDM, SDM, TDM.)

NAK A negative acknowledgement. (*See* ACK.)

NEPRA Acronym for New England Packet Radio Association.

network An interconnection of computer systems, terminals, and communications facilities.

network layer Level 3 of the OSI/RM. Deals with addressing, routing, multiplexing, and flow control. Two types of networks are virtual circuit and datagram.

network node Also called packet switch. A hardware system with a level 3 protocol designed to forward packets through the network to their destination. (*See* datagram, virtual circuit.)

node A general term used to indicate the different stations in a packet network. Nodes may be terminal nodes, network nodes, station nodes, and others. (*See* TNC, digipeater.)

NRZ Acronym for NonReturn to Zero. An encoding technique for binary digital signals in which a binary "1" is encoded as a positive pulse and a binary "0" as a negative pulse. When modulated, the positive pulse becomes the mark tone and the negative pulse the space tone. This is the encoding technique used in Baudot RTTY. (*See* Manchester.)

null A "blank"; a meaningless character usually used to consume extra bit space or time.

null modem Also null modem adaptor and null modem cable. An RS-232 interfacing device which switches several pins to allow DTEs or DCEs to communicate with devices of the same designation.

nybble A group of four bits. One half of a byte. Represented by a single HEX character.

NZRI Acronym for NonReturn to Zero Inverted. An encoding technique for binary digital signals in which a binary "0" causes a change in signal level while a binary "1" causes no change. This is the encoding technique in use in most amateur packet radio systems. (*See* Manchester.)

octal A number system based on powers of 8. Characters are 0-7. (*See* HEX, binary.)

octet A group of eight bits. (*See* byte, nybble.)

OSCAR Acronym for Orbital Satellite Carrying Amateur Radio.

OSI/RM Acronym for Open Systems Interconnection Reference Model. A formal hierarchical identification of all network functions as established by the ISO.

overhead Any information other than the actual data that is transmitted. In packet radio, some overhead is necessary (i.e., addresses, control information), but it should be kept to a minimum.

PAC-COMM An abbreviation for Pac-Comm Packet Radio Systems, Inc.

packet A group of bits including data and control elements which is transmitted as a whole. Technically, a packet is not formed until the network layer; however, many refer to frame transmissions on the data link layer as packets. (*See* frame.)

packet controller Term used for a hardware TNC with an on-board modem.

packet switch A device which is used along with a network layer protocol to forward (switch) data sent on the network to the next node. Packet switches acknowledge data sent to them and then wait for an acknowledgement from the next node. In most cases, individual users are not involved in selecting the routing used by the packet switches. (*See* network node.)

PACSAT Contraction of PACket SATellite. A packet satellite designed by AMSAT which will feature a flying mailbox similar to terrestrial BBSs. (*See* JAS-1, RUDAK.)

PAD Acronym for Packet Assembler/Disassembler. May be used interchangeably with TNC. (*See* FAD.)

padding The addition of an extra character to a group of characters in order to reach a predefined amount.

parallel transmission A method of transmitting data in which all bits of each bit grouping are transmitted simultaneously on separate channels. (*See* serial transmission.)

parameter A variable stored for future reference.

parity The addition of a noninformation bit to a group of bits making the total number of binary "1"s in the group either even or odd depending on the type of parity selected. This permits single bit error detection in each group.

path The sequence of channels, gateways, and repeaters used to transmit information from one node to another.

peripheral Any device which can be connected to a computer system to extend its operating capabilities.

physical layer Level 1 of the OSI/RM. Concerned with electrical characteristics of the communications link. (*See* modem, RS-232C.)

PLL Acronym for Phase Locked Loop. A circuit for synchronizing an oscillator with the phase of a signal.

polling system A method of TDM (Time Division Multiplexing) in which each station is asked (polled) to determine if it has any traffic to send. (*See* random access.)

PPRS Acronym for Pacific Packet Radio Society.

presentation layer Level 6 of the OSI/RM. Performs any code conversion, handles control data structure and display formats. Also manages data interchange with peripheral storage devices.

PRM Acronym for *Packet Radio Magazine*. A monthly publication formerly published by FADCA.

propagation delay The time lapse between transmission and reception of a signal on a radio link. Satellite delays tend to be longer than those of terrestrial links.

protocol A formal set of rules which dictate the format, timing, and other parameters of message exchange between two or more devices.

prototype A very early design of a product assembled in small quantities for initial testing and further development. Sequence is: conception, design, prototype, alpha testing, beta testing, and production.

PSK Acronym for Phase Shift Keying. A method of transmitting digital information in which the phase of the carrier is varied in accordance with the digital signal.

PSR Acronym for Packet Status Register. The newsletter published by TAPR as part of PRM.

RAM Acronym for Random Access Memory. Electronic memory which may be read from and written to. However, once power is removed, all stored data is lost. Some RAMs provide for battery backup to retain the data in case normal power is removed.

random access A type of network in which stations may transmit at any time provided the channel is available. (*See* polling system.)

ring network A network configuration in which each node is connected to two other adjacent nodes, one on each side. When the connections are complete, the path of connections will resemble a ring or circle. Each node may communicate directly only with the node immediately preceding and immediately following it. All nodes serve as relay stations to allow for communications throughout the network. (*See* star, bus.)

ROM Acronym for Read Only Memory. Electronic memory which may be read from but not written to. Data is permanently retained. Some ROMs allow for occasional programming and erasure (the EPROM—Erasable Programmable ROM). Most ROMs can be erased with UV light, so the top of the ROM chip is covered with a label or sticker to block all light.

RS-232 1. An EIA standard. The latest version is D. A common serial communications interface for computer peripherals. 2. Defines the voltage signaling levels in electronic equipment. Range is -25 to -5 and $+5$ to $+25$ volts; $+/-12$ and $+/-15$ volts are commonly used.

RTS Acronym for Request To Send in the RS-232C standard. Also referred to as Pin 4, CA by the EIA, and 105 by the CCITT.

RTTY Contraction of Radio TeleTYpe. Direct printing digital radio communications.

RUDAK A packet experiment designed by the West German affiliate of AMSAT included on the Phase IIIC satellite. (*See* JAS-1, Pacsat.)

SDLC Acronym for Synchronous Data Link Control. An IBM data link protocol very similar to HDLC.

SDM Acronym for Space Division Multiplexing. A method of allowing multiple users to share a single communications channel by arranging the users so that they are not in each other's communications range. (*See* FDM, TDM.)

serial transmission A method of transmitting data in which each bit is sent sequentially on a single channel. (*See* parallel transmission.)

session layer Level 5 of the OSI/RM. Initiates and terminates communications through the network; also handles network log-on and authentication.

simplex Operation over a single channel in one direction at a time. (*See* full duplex, half duplex.)

smart terminal A communications terminal with advanced capabilities (such as Xon/Xoff flow control, buffers, variable parameters, and echoing) in addition to those required for basic communications. (*See* dumb terminal.)

software The program and procedures that control the operation of hardware systems.

standard An established procedure, model, or design that has gained widespread recognition and conformity. Can be developed by committee, industry, or popular usage.

star network A network in which all user nodes are situated about a central node. All communications between user nodes must take place through the central node. (*See* ring, bus.)

start bit Used in asynchronous serial communications. The first bit in each character which notifies the receiver that a bit group is coming.

stop bit Used in asynchronous serial communications. The last bit in each bit group which notifies the receiver that the bit group is ended.

synchronous Transmission in which the data is transmitted at a fixed rate with the transmitter and receiver synchronized. This eliminates the need for start and stop elements used with asynchronous transmission. (*See* asynchronous.)

TAPR Acronym for Tucson Amateur Packet Radio. A nonprofit organization specializing in packet radio development.

TDM Acronym for Time Division Multiplexing. The sharing of a single communications channel between many users by allotting the channel to each user on a time basis. (*See* multiplexing, FDM, SDM.)

teleport A gateway between a terrestrial station or network and a satellite.

teletype A registered trademark of the Teletype Corporation. Generic term used is teleprinter. A typewriter-like device with a mechanical system to change keypresses into electrical pulses for transmission. Received pulses are converted back to characters which are printed on paper.

terminal A dedicated communications device that usually has a keyboard, display device(s), and an I/O port. (*See* dumb terminal, smart terminal.)

throughput The actual rate of transmission (usually in BPS) taking into account switching times, retransmissions, and other delays.

TNC Acronym for Terminal Node Controller. A device which assembles and disassembles frames. Usually includes some form of a user interface and command set. May be implemented in hardware or software. Used in conjunction with a radio, modem, and terminal for packet radio applications. (*See* node, PAD, FAD, packet controller.)

toggle To switch between one of two possible conditions.

token passing A form of TDM in which a unique binary sequence is passed from node to node. Only the node with the token may transmit.

transponder A device that receives radio signals in one segment of the frequency spectrum and repeats them on another segment of the spectrum.

transport layer Level 4 of the OSI/RM. Arranges data in order in the event packets arrive out of order.

TTL Acronym for Transistor Transistor Logic. A logic standard which represents a binary "1" as +5 volts and a binary "0" as 0 volts. TTL logic is used in the internal operations of most computer circuits.

turnaround delay The time period required for a station to switch between receive mode and transmit mode.

UART Acronym for Universal Asynchronous Receiver Transmitter. A device, usually packaged as an IC, which transmits and receives asynchronous serial data. The transmitter accepts data in parallel format and outputs the data in serial format. The receiver accepts data in serial format and outputs the data in parallel format.

unipolar keying A technique in which a binary "1" is represented by a pulse and a binary "0" by the absence of a pulse. Unipolar keying's poor performance led to the development of bipolar keying.

uplink A radio link originating at a ground station and terminating at a satellite. (*See* downlink.)

user interface The interface between the user and the device being used. In packet radio, the procedures implemented to allow the user to communicate with the TNC. Usually consists of a computer program incorporating a variety of commands. (*See* command set.)

VADCG Acryonym for Vancouver Amateur Digital Communications Group.

VADCG Protocol An early level 2 protocol based on HDLC developed by Doug Lockhart VE7APU for use in packet radio development. No longer in widespread use. Now called the V-1 protocol. (*See* AX.25.)

VDT Acronym for Video Display Terminal. A terminal with a video monitor.

virtual circuit A type of packet networking in which a logical connection is established prior to the transfer of data. This allows for abbreviated addressing and lower overhead at the expense of routing flexibility. (*See* datagram.)

X.25 A CCITT standard. "Interface Between DTE and DCE for Terminals Operating in the Packet Mode on Public Data Networks." Defines the architecture of three levels of protocols existing in the serial interface between a packet mode terminal and a gateway to a packet network. AX.25 level 2 is derived from the data link layer of X.25.

X.28 A CCITT standard. "DTE/DCE Interface for a Start-Stop Mode DTE accessing the PAD in a Public Data Network." Defines the protocol structure in a serial interface between an asynchronous terminal and an X.3 PAD.

X.29 A CCITT standard. "Procedures for the Exchange of Control Information and User Data Between a PAD and a Packet Mode Terminal DTE or another PAD." Defines the protocol structure between two PADs or between a PAD and a Packet Mode Terminal.

X.3 A CCITT standard. "Packet Assembly/Disassembly Facility (PAD) in a Public Data Network." Describes the PAD which normally is used at a network gateway to allow connection of an asynchronous terminal to a packet network.

X.121 A CCITT standard. "International Numbering Plan for Public Data Networks." Defines an addressing plan with code assignments for each nation.

Xerox 820 A single board Z-80 based microcomputer. Usually equipped with 64K memory, a parallel printer port, and two RS-232 serial ports. Often combined with a keyboard, monitor, and disk drives to form a complete system. The Xerox 820 is used in many amateur packet radio applications including the W0RLI BBS, dual-port digipeaters, and dedicated TNCs.

Bibliography

Over the years many books and articles have been written on the subject of packet switching, packet radio, and amateur packet radio. To attempt to list them all here would be impractical for several reasons. First of all, it would take a great deal of time and effort to accumulate the necessary listing information. Second, the amount of space such a list would consume is out of proportion to this book. Third, and most important, such a list would be of little help. With so many listings, it would be difficult if not impossible to select those works which would be of most interest to you without much time spent locating and examining them.

This section of the book contains a sizable listing of books, proceedings, papers, and articles dealing with amateur packet radio in some way. I have attempted to select those works that reinforce and expand upon the concepts that are presented in this book. Of course, the standard references (as established by members of the packet radio community) are also included.

BOOKS AND ANTHOLOGIES
ARRL. *ARRL Amateur Radio Computer Networking Conferences 1-4.* $18 from ARRL.

———. *The ARRL Handbook for the Radio Amateur.* Newington: ARRL, 1985 and 1986.

Davidoff, Martin K2UBC. *The Satellite Experimenter's Handbook.* Newington: ARRL, 1984.

Folts, Harold C. and Harry R. Karp, ED. *Data Communications Standards.* New York: McGraw-Hill, 1979.

Friend, George E., et al. *Understanding Data Communications.* Dallas: Texas Instruments, 1984. (Also available from Radio Shack; Catalog No. 62-1389)

Green, Jr. Paul E., Ed. *Computer Network Architectures and Protocols.* New York: Plenum Press, 1982.

Grubbs, Jim K9EI. *Get***CONNECTED to Packet Radio.* Springfield: Q-Sky, 1986.

_____. *The Digital Novice.* Springfield: Q-Sky, 1987.

Horzepa, Stan WA1LOU. *Your Gateway to Packet Radio.* Newington: ARRL, 1987.

Lancaster, Don. *TV Typewriter Cookbook.* Indianapolis: Howard W. Sams and Co., 1976.

McNamara, John E. *Technical Aspects of Data Communication.* Bedford: Digital Press, 1982.

Meijer, Anton, and Paul Peters. *Computer Network Architectures.* Rockville: Computer Science Press, 1982.

Rouleau, Robert VE2PY, and Ian Hodgson VE2BEN. *Packet Radio.* Blue Ridge Summit: TAB Books, 1981.

Seyer, Martin D. *RS-232 Made Easy: Connecting computers, printers, terminals, and modems.* Englewood Cliffs: Prentice-Hall, 1984.

Tanenbaum, Andrew S. *Computer Networks.* Englewood Cliffs: Prentice-Hall, 1981.

TAPR. *TAPR Packet Radio TNC System Manual.* 1984.

ARTICLES AND PAPERS

Adams, W. Max W5PFG. "Briefly Speaking: Basic Amateur Radio Packet Radio." *CQ.* November 1985: 13-20.

Beeler, Michael. "Degradable Performance in Packet Switching Networks." *Proceedings of the IEEE Computer Society COMPCON 82 Fall: Computer Networks.* IEEE Catalog No. 82CH1796-2. (1982): 437-443.

Bishop, Jeffrey N7FDS. "Seeing Packet Radio With Different Eyes." *73.* August 1986: 48-49.

Boorstyn, Robert R., et al. "A New Acknowledgement Protocol for Analysis of Multihop Packet Radio Networks." *Proceedings of the IEEE Computer Society COMPCON 82 Fall: Computer Networks.* IEEE Catalog No. 82CH1796-2. (1982): 383-393.

Borden, David W. K8MMO, and Paul L. Rinaldo W4RI. "The Making of an Amateur Packet-Radio Network." *QST*. October 1981: 28-30.

Churchward, Budd WB7FHC. "Packet Radio: Getting Started." *World Radio*. April 1986: 6+.

Davey, J.R. "Modems." *Proceedings of the IEEE*. Vol 60 1972. 1284-1292.

Flammer, George WB6RAL. "Survival Training for Mountaintop Digipeaters." *73*. August 1986: 68-73.

Hodgson, Ian VE2BEN. "An Introduction to Packet Radio." *Ham Radio*. June 1979: 64-67.

Hutton, Louis K7YZZ. "Connect Alarm!" *73*. August 1986: 66.

Johnson, Lyle WA7GXD. "Join the Packet Radio Revolution." *73*. September 1983: 19-24.

_____. "Join the Packet Radio Revolution—Part II." *73*. October 1983: 20-31.

_____. "Join the Packet Radio Revolution—Part III." *73*. January 1984: 36-44.

Karn, Phil KA9Q. "Beyond Level Two." *73*. August 1986: 74-78.

Langner, John WB2OSZ. "Precision Packet Tuning." *73*. August 1986: 40-47.

Macassey, Julian N6ARE. "Computers and Hams." *World Radio*. February 1986: 26-27.

Markoff, John. "Bulletin Boards in Space." *Byte*. May 1984: 88-94.

Mayo, Jonathan L. KR3T. "An Amateur Packet Radio Primer: Part I - Introduction." *CQ*. November 1986: 11-17.

_____. "An Amateur Packet Radio Primer: Part II - Packet Radio Equipment." *CQ*. December 1986: 55-60.

_____. "An Amateur Packet Radio Primer: Part III - Operating Packet Radio." *CQ*. January 1987: 18-20.

_____. "An Amateur Packet Radio Primer: Part IV - Bulletin Board Operation." *CQ*. February 1987: 30-33.

_____. "Amateur Packet Radio. Who Needs It? You Do!" *CQ*. November 1988: 11-17.

_____. "Amateur Packet Radio Networking and Protocols, Part 1." *Ham Radio*. February 1988: 33-38.

_____. "Amateur Packet Radio Networking and Protocols, Part 2." *Ham Radio*. March 1988: 56-64.

_____. "Amateur Packet Radio Networking and Protocols, Part 3." *Ham Radio*. April 1988: 41-45.

———. "Understanding the RS-232 Standard." *CQ*. November 1988: 32-36.

Morrison, Margaret KV7D, and Dan Morrison KV7B. "Amateur Packet Radio: Part 1." *Ham Radio*. July 1983: 14-18.

Morrison, Margaret KV7D, and Dan Morrison KV7B, and Lyle Johnson WA7GXD. "Amateur Packet Radio: Part 2." *Ham Radio*. August 1983: 18-29.

Pearce, Jon WB2MNF. "So You Want To Be A Sysop?" *73*. August 1986: 50-55.

Popiel, Glen WA4FTX. "Packet Radio Operating Tips." *World Radio*. April 1986: 28.

Price, Harold NK6K. "What's all this Racket about Packet?" *QST*. July 1985: 14-17.

———. "A Closer Look at Packet Radio." *QST*. August 1985: 17-20.

———. "et al. "Birds 'N' Bauds." *73*. August 1986: 58-64.

———. "And If That Wasn't Enough . . ." *73*. August 1986: 80-85.

Reedy, Gwyn W1BEL. "A Packet Primer." *73*. August 1986: 28-32.

Rouleau, Robert T. VE2PY. "The Packet Radio Revolution." *73*. December 1978: 192-193.

Sternberg, Norm W2JUP. "How To Make Friends at 1200 Baud." *73*. August 1986: 34-39.

Zorpette, Glenn. "The High-Tech Hobbyhorse." *IEEE Spectrum*. May 1985: 96-98.

About the Author

Jonathan L. Mayo has been involved with electronics, computers, and digital communications for many years. He is a licensed Extra class amateur radio operator, callsign KR3T, and has been writing professionally for several years in science and technology subject areas such as radio-based digital communications, electronics technology, and general science.

Jon has written numerous articles dealing with amateur radio—from equipment reviews to feature articles. He is one of the most prolific authors of amateur packet radio material. Some of his articles have been translated and published in foreign amateur radio magazines. He also has given many presentations and demonstrations on digital communications for amateur radio clubs and other organizations.

Index

Index

A
ABAUD, 109
ABIT, 109
acknowledgement (ACK), 11
  end to end, 127
additional text message, 123-125
address fields, 69
address flag, 67
AEA, 142
AEA PKT-1 TNC, 28
AFTER#, 117
ALOHANET, 21, 22
Alpha TNC, 25, 26
amateur radio
  call signs for, 199
  licenses for, 198
  packet radio operation by, 199
Amateur Radio Computer Networking Conferences, 24, 27, 28
Amateur Radio Emergency Communications Organization (USSR), 3
Amateur Radio Research and Development Corp., 12, 24
amateur radio, 197-199
amateur teletype over radio (AMTOR), 9, 10
amplifiers, 103
amplitude modulation, 43
AMRAD, 26
AMSAT, 26, 167, 168
AND, 15
antennas, 49, 103-104
Apple IIe computers, 93
application layer, 63
application programs, 18
ARRL, 3, 24
ARRL Handbook, 80
ART-1 interface, 155
ASCII code, 8, 24, 175-176
ASCII RTTY, 9, 10
asynchronous input-output, TNCs and, 36

audio frequency shift keying (AFSK), 43, 45
AUTOLF, 111
AWLEN, 109
AX.25 protocol, 12, 26, 28, 33, 39, 62, 68-74, 76
  components of, 69
  tracing, 72
  versions of, 69
AXDELAY, 113

B
Baker, Mark, 25
bandwidth, 44
BASIC, 19
baud rates, 21, 43, 88
  channel characteristics and, 44
Baudot code, 8, 9
Baudot RTTY, 9, 10
BBS operation, 28, 31, 128-137, 140
  command summary for, 136
  file system, 133-136
  future development of, 171
  gateway system, 136
  mail system, 130-133
  WORLI, 28, 31
BEACON, 117
beacons, 117-118, 139
Bell 103 modem, 63
Bell 202 modem, 47, 48, 63
Beta TNC, 26, 27, 28
Bill Ashby and Sons, 146
binary code, 6, 7, 8, 15
bipolar keying, 63
bit oriented protocols, 23, 66
bit stream, 40
bit stuffing, 66
bits, 6
bits per second (BPS), 22, 44, 109
BKONDEL, 112

223

BKSP, 111
Boolean algebra, 15
British National Physics Laboratory, 20
budget station, 96-98
bulletin board packet systems (see BBS operation)
bursts, 10
Busch, Mike, 78
busy message, 120

C

cable-based networks, 20
CALL, 126
call signs, 13, 70
   amateur radio, 199
Canadian packet radio, 22, 77
CANLINE, 112
CANPAC, 112
carrier sense system, 91
carrier-sense multiple access with collision detection (CSMA-CD), 10, 55
cathode ray tubes (CRT), 16
CCITT X.25 protocol, 26
central processing units (CPU), 16, 39
Chamberlain, Marc, 25
channel characteristics, 44
character oriented protocol, 66
Clark, Jerry, 25
Clark, Tom, 26
clock, 39
cloned TNC, 28
CMD, 118, 119, 121
CMOS technology, 41
COBOL, 19
collision detection, 11
COMMAND, 112, 118, 121
Commodore computers, 94, 97
Compaq computers, 93
Computer Networks, 22, 80
CONMODE, 118, 121
CONNECT, 118, 138
connected message, 121, 122
Connors, Den, 25
CONOK, 114, 118, 119
contesting, 173
continuous wave (CW), 9, 197
CONVERSE, 114, 117, 121
COP-BOP protocol, 77
count, 68
CP-M, 18, 134
CSMA-CD protocol, 22
current loop input-output, 89
cyclic redundancy check (CRC), 68

D

DARPA, 20
data communications equipment (DCE), 178
data link layer, 62, 65
data terminal equipment (DTE), 178
datagram, 79

Davidoff, Martin, 167
Davies, D.W., 20
Dayton Hamvention, 28
dedicated terminal, 83
DELETE, 111
delimited frame, 62
demodulation, 47
differential phase shift keying (DPSK), 22
DIGIPEAT, 114
digipeaters, 5, 13, 24, 126-128, 139
   monitor connection through, 126-128
digital communications, 6
   packet radio vs., 9
digital radios, 166
DISC, 71
DISCONNECT, 118, 124
disconnect message, 125, 126
disconnect request (DISC), 71
disconnected mode (DM), 71
discrete event, 44
disk drive, 17
disk operating system (DOS), 18, 134
DM, 71
downloading, 87
DR-100/200, 158
DRSI, 142, 146
dual port digipeaters, 127
DWAIT, 115

E

EASTNET, 27
EBCDIC code, 8
ECHO, 111
echoing, 87
emulator, terminal, 83
encoding techniques, 63
end to end ACK, 127
equipment and accessories, 142-164
error checking, 11

F

FADCABEACON, 26
Federal Communications Commission (FCC), 24
file system BBS operation, 133-136
file transfer protocol (FTP), 80
filenames, 134
filtering, 40
firmware, 18
flags, 40, 69
Florida Amateur Digital Communications Assoc. (FADCA), 26, 77
FLOW, 114
flow control, RS-232 and, 183
FM radio, 197
FORTRAN, 19
FRACK, 115
frame assembler-disassembler (FAD), 34
frame check sequence (FCS), 11, 39, 42, 67
frame reject (FRMR), 71
frames, 11, 67

frequencies, 11, 195-196
frequency division multiplexing (FDM), 52
frequency modulation, 43
frequency shift keying (FSK), 43, 45
FRMR, 71
FUGI-OSCAR-12, 167
full duplex, 9

G
gates, 15
Gateway, 136
gateway stations, HF, 48
gateway system BBS operation, 136
gateways, 129
GLB Electronics, 28, 142, 147

H
HAL, 150
half duplex, 9
Ham Radio Magazine, 22, 27, 77
Hamilton Area Packet Network, 150
HAPN Packet Adapter TNC, 150
HBAUD, 114
HD-4040 TNC, 152
HDLC protocol, 27, 39, 42, 62, 65-68
Heathkit TNC, 29, 142, 150
Henderson, Dave, 27
HF band, 11, 46, 47, 195, 196
 operating tips for, 140
HF*Modem, 147
high-level data link control (HDLC) format, 11, 62
high-speed modems, 165
HK-21 TNC, 151
HK-232, 150

I
IBM PC computers, 93
identification, 13
initial parameters, 108-116
input-output
 devices, 18
 radio, 40
insulation displacement, 186
integrated circuits, 15-16
interface, 69
 ART-1, 155
international standards organization (ISO), 69
internet protocol, 79
intranet protocol, 79

J
JAS-1, 167, 168
Johnson, Lyle, 25, 28, 31

K
KAM TNC, 152
Kantronics, 29, 142, 152

Karn, Phil, 80
Kenwood transceiver, 101
KPC-2 TNC, 153
KPC-2400 modem, 154
KPC-4 TNC, 153

L
languages, computer, 18
LCOK, 111
level 2 protocols, 76
licenses, amateur radio, 198
link access procedure balanced (LAPB), 70
local area networks (LANs), 172
Lockhart, Doug, 28, 74

M
machine code, 18
Macket program, 160
MacPacket program, 160
mail system BBS operation, 130-133
mark, 8
MAXFRAME, 116
McClain, Dave, 25
memory, 16
 TNC, 38
menehune, 21
MFJ, 142, 154
microcomputer systems, 16
microlog, 155
microphones, 92
Micropower-2 TNC, 157
modem-radio interface, 91
modems, 8, 11, 22, 42, 92, 144
 2400 TNC, 154
 amateur packet radio schemes for, 46
 Bell 103, 63
 Bell 202, 47, 48
 HF*, 147
 high speed, 165
 KPC-2400, 154
 Netlink 220, 149
 null, RS-232 and, 182
 shift filter, 47
 ST-7000, 150
modulation, 45
 amateur packet radio schemes for, 46
 digital, 43
 pulse duration (PDM), 65
 pulse position (PPM), 65
monitor, 18, 116
monitoring, 116-128
 beacons in, 117
 connection through digipeaters, 126-128
 connection to another station, 118
 terminal connection, 118
Montreal Amateur Radio Club, 22
Moran, Bill, 24
Morrison, Margaret, 27
Morse code, 197
MOS technology, 41
MS-DOS, 18

multiconnect capability, 124
multiplexing, 51
Murray code, 8, 9
MYVADR, 115

N
NAND, 15
National Aeronautics and Space Administration (NASA), 168
national traffic system (NTS), 131
net control station (NCS), 59
NET-ROM operations, 137-139, 160
Netlink 220 modem, 149
network layer, 62, 77
network node controller (NNC), 79
networks, 11, 50
  cable-based, 20
  future development of, 171
NNC TNC, 162
node, 21
non-return to zero (NRZ), 63
NOR, 15
North American Presentation Level Protocol Syntax (NAPLPS), 80
NRZ inverted (NRZI), 64
NRZ space (NRZ-S), 64
NRZ-Mark (NRZ-M), 65
null modem, 182
NULLS, 111
Nyquist's equation, 44, 45

O
1270/74/78/73 series TNC, 154
octet, 70
open systems interconnection reference model (OSI-RM), 69
operating systems, 18
operational parameters, 113-116
OR, 15
Oredson, Hank, 31, 128
organizations, 188-191
OS-9, 18
OSCAR 13, 167

P
Pac-Comm, 142, 156
PAC-NET board, 146
packet assembler-disassembler (PAD), 34
Packet Communicator TNC, 29
Packet Radio, 22, 26, 77
packet radio
  amateur organizations for, 22
  applications for, 2
  contests, 173
  CSMA-CD networking for, 10
  development of, 1-33
  digital communications basics for, 6
  early cable-based networks, 20
  equipment and accessories, 142-164
  future of, 165-174
  hardware systems, 34-49
  history of, 20-33
  initial cost of, 15
  initial parameters for, 108-116
  integrated circuits in, 15
  introduction to, 5
  microcomputer systems for, 16
  monitoring, 116-128
  multiplexing model for, 56
  NET-ROM operation for, 137-139
  network schematic of, 5
  networking and protocols, 50-81
  operating tips for, 139-140
  operation of, 105-141
  organizations, 188-191
  other digital modes vs., 9
  satellite communications for, 167
  station layout, 13
  station setup, 82-104
  system trace, 41
  user interface for, 106-108
Packet Status Register, 25
packet status register (PSR), 25
Packet, The, 24, 76, 77
PACLEN, 116
PACSAT, 168
Pakratt 64 TNC, 145
parallel communication, 37, 38
parameters
  operational, 113-116
  radio, 113
  terminal, 109-113
PARITY, 110
PC-120 TNC, 158
PC-PACKET program, 160
PC*Packet Adapter TNC, 146
Phase III-C satellite, 167
phase modulation, 43
phase shift keying (PSK), 43, 46
physical layer, 63
PK-1 TNC, 28, 147
PK-1L TNC, 148
PK-232 data controller, 143
PK-88 TNC, 143
PKT-1 TNC, 144
PM-1 packet modem, 144
polling, 54
portable packet radio station, 98-101
presentation layer, 63, 80
Price, Harold, 27
printer, 17
ProDOS, 18
programmable control unit (PCU), 22
programmable read only memory (PROM), 38
protocol, 11, 12
  AX.25, 26, 33, 39, 62, 68-74, 76
  bit-oriented, 23, 66
  CCITT X.25, 26
  character oriented, 66
  COP-POB, 77
  CSMA-CD, 22
  data link layer, 65
  file transfer, 80

function of, 57
future development in, 169, 170
HDLC, 23, 39, 42, 62, 65-68
implementation programs, 39
internet, 79
intranet, 79
level 2, 76
network layer, 77
North American Presentation Level, 80
physical layer, 63
presentation layer, 80
session layer, 80
simple mail transfer, 80
SLDC, 76
TAPR-DA, 77
TCP-IP, 80
transmission control (TC), 80
transport layer, 80
V-1, 33, 74-76
V-2, 33, 76
VADCG, 39, 74-76
protocol identifier field (PID), 70
pulse duration modulation (PDM), 65
pulse position modulation (PPM), 65

Q
QSO, 3
QST Magazine, 24, 29
quaternary amplitude modulation (QAM), 46

R
radio (see also transceivers), 48
  digital, 166
  parameters for, 113
  selection of, 90-93
Radio Amateur Telecommunications Society (RATS), 12, 26, 77
radio frequency interference (RFI), 101-102
radio input-output, 40
radio teletype (RTTY), 6, 197
Raikes, Ron, 78, 138, 162
Rand Corporation, 20
random access, 54, 55
random access memory (RAM), 16, 38
read only memory (ROM), 16, 38
receive not ready (RNR), 72
receive ready (RR), 72
received frames message, 122
REDISPLA, 112
Reedy, Gwyn, 32
reject (REJ), 72
repeaters, 13
  DR-100/200, 158
RETRY, 115, 119, 121
retry out, 119, 120
RFM-220 radio modem, 144
Richardson, Robert, 160
Richcraft Engineering, 96, 160
RNR, 72
RPC-2000 TNC, 150
RR, 72

RS-232 standard, 36, 37, 63, 88, 89, 101, 177-187
  flow control, 183
  full cable configuration for, 182
  introduction to, 178
  limitations to, 185
  minimum cable configuration for, 181
  nonstandard implementation of, 184
  null modem for, 182
  signal levels for, 181
  signal pins for, 179
  wiring cables for, 185
RUDAK, 167

S
73 Magazine, 77
S. Fine Software, 160
SABM, 71
SAREX 2, 168
satellite communications, 48, 167
Satellite Experimenters Handbook, The, 167
SENDPAC, 112
sequence number, 68
serial communication, 36, 37, 40
session layer, 62, 80
set asynchronous balanced mode (SABM), 71
Shannon's Law, 45
shared terminal packet radio station, 98
shift filter-based modem, 47
shuttle amateur radio experiment (SAREX), 168
simple mail transfer protocol (SMTP), 80
simplex, 9
simplex packet repeaters, 13
single side band (SSB), 197
slave station, 70
SLDC protocol, 76
software, 18, 169
Software 2000, 160
SOUTHNET, 27
space, 8
space division multiplexing (SDM), 55
space shuttle, 168
speakers, 92
squelch detect lead, 91
ST-7000 modem, 150
standards, 69
  organizations for, 191-192
static FDM, 54
station layout, 13
station node controller (SNC), 23
station setup, 82-104
  antennas for, 103-104
  budget station, 96-97
  portable station, 98-101
  radio frequency interference and, 101
  shared terminal, 98
  typical station, 93-96
sub-station identifier (SSID), 70
surface-mount TNC, 158
synchronous communications, 40
SYSOP BBS operation, 129

T

2400 TNC modem, 154
T-R time, 48
TAPR, 27, 28, 161
TAPR TNC-1 System Manual, 74, 76
TAPR-DA procotol, 77
teleport, 5
television, 197
terminal control unit (TCU), 22
terminal emulator, 83
terminal node controller (TNC), 11, 13, 23, 34
  1270/73/74/78 series, 154
  AEA PKT-1, 28
  Alpha, 25
  asynchronous input-output, 36
  Beta, 26
  clones, 28
  HAPN Packet Adapter, 150
  hardware-based, 36
  HD-4040, 152
  Heathkit, 29
  HK-21, 151
  KAM, 152
  kits and board, 28
  KPC-2/4, 153
  memory, 38
  Micropower-2, 157
  NNC TNC, 162
  Packet Communicator, 29
  Pakratt 64, 145
  PC-120, 158
  PC*Packet Adapter, 146
  PK-1, 28, 147
  PK-1L, 148
  PK-88, 143
  PKT-1 clone, 144
  power requirements for, 41
  radio input-output, 40
  RPC-2000, 150
  selection of, 82, 89-90
  software based, 34, 169
  surface mounted, 158
  system trace for, 41
  TINY-2, 157
  TNC Plus, 33, 162
  TNC-1, 36, 161
  TNC-2, 29, 161
  TNC-200, 157
  TNC-220, 156
  TNC-2A, 150
  TRS-80 Models, 35
terminals, 11
  connection of, 118
  parameters for, 109-113
  selection of, 82-89
  shared, 98
test sent-received message, 123
time division multiplexing (TDM), 52, 54
TINY-2 TNC, 157
TNC Plus, 33, 162
TNC-1 TNC, 36, 161
TNC-2, 29, 161
TNC-200, 157
TNC-220, 156
TNC-2A, 150
token pasing, 54
TRANS, 114
transceivers, 14, 48
  digital, 166
  selection of, 90-93
transistors, 15
transmission control protocol (TCP), 80
transmission control protocol-internet protocol (TCP-IP), 80
transmission medium, 8
transport layer, 62, 80
TRS-80 computers, 93, 96, 98
TRSDOS, 18
Tucson Amateur Packet Radio Corp. (TAPR), 25
TV Typewriter Cookbook, 89
TXDELAY, 113, 115
typical packet radio station, 93-96

U

UA, 71
UI, 72
unbalanced ground, 185
United States packet radio organizations, 24
universal asynchronous receiver-transmitter (UART), 38
UNIX, 18
unnumbered acknowledgement (UA), 71
unnumbered information (UI), 72
UNPROTO, 117
uploading, 87
user interface, 106-108
  future development in, 170

V

V-1 protocol, 33, 74-76
V-2 protocol, 33, 76
VADCG protocol, 28, 33, 39, 74-76
Vancouver Amateur Digital Communication Group (VADCG), 23, 142, 162
VDIGIPEA, 114
VHF band, 11, 46, 47, 195, 196
virtual circuits, 79, 80
voltage controlled oscillator (VCO), 47

W

WA8DED software, 162
WESTNET, 27
wildcard characters, 135
WORLI packet bulletin board, 28, 31, 128

X

XFLOW, 112
XMITOK, 115, 118
XOR, 15

Other Bestsellers From TAB

☐ **SUPERCONDUCTIVITY—THE THRESHOLD OF A NEW TECHNOLOGY—Jonathan L. Mayo**

Superconductivity is generating an excitement in the scientific world not seen for decades! Experts are predicting advances in state-of-the-art technology that will make most existing electrical and electronic technologies obsolete! This book is one of the most complete and thorough introductions to a multifaceted phenomenon that covers the full spectrum of superconductivity and superconductive technology. 160 pp., 58 illus.
**Paper $14.95   Hard $18.95
Book No. 3022**

☐ **FIBEROPTICS AND LASER HANDBOOK—2nd Ed.—Edward L. Safford, Jr. and John A. McCann**

Explore the dramatic impact that lasers and fiberoptics have on our daily lives—PLUS, exciting ideas for your own experiments! Now, with the help of experts Safford and McCann, you'll discover the most current concepts, practices, and applications of fiberoptics, lasers, and electromagnetic radiation technology. Included are terms and definitions, discussions of the types and operations of current systems, and amazingly simple experiments you can conduct! 240 pp., 108 illus.
**Paper $19.95   Hard $24.95
Book No. 2981**

☐ **101 SOLDERLESS BREADBOARDING PROJECTS—Delton T. Horn**

Would you like to build your own electronic circuits but can't find projects that allow for creative experimentation? Want to do more than just duplicate someone else's ideas? In anticipation of your needs, Delton T. Horn has put together the ideal project *ideas* book! It gives you the option of customizing each project. With over 100 circuits and circuit variations, you can design and build practical, useful devices from scratch! 220 pp., 273 illus.
**Paper $18.95   Hard $24.95
Book No. 2985**

☐ **ELECTRONIC DATABOOK—4th Edition—Rudolf F. Graf**

If it's electronic, it's here—current, detailed, and comprehensive! Use this book to broaden your electronics information base. Revised and expanded to include all up-to-date information, the fourth edition of *Electronic Databook* will make any electronic job easier and less time-consuming. This edition includes information that will aid in the design of local area networks, computer interfacing structure, and more! 528 pp., 131 illus.
**Paper $25.95   Hard $34.95
Book No. 2958**

Other Bestsellers From TAB

☐ **HOW TO MAKE PRINTED CIRCUIT BOARDS, WITH 17 PROJECTS—Calvin Graf**

Now *you* can achieve the polished look of skillfully etched and soldered boards with the help of Calvin Graf. This book explains thoroughly and *in plain English*, everything you need to know to make printed circuit boards (PCBs). Key subjects include: getting from an electronic schematic to a PCB, etching a printed circuit board, cleaning, drilling, and mounting electronic parts, soldering, and desoldering project kits. 224 pp., 177 illus.
**Paper $18.95   Hard $23.95
Book No. 2898**

☐ **BEYOND THE TRANSISTOR: 133 ELECTRONICS PROJECTS—Rufus P. Turner and Brinton L. Rutherford**

Strongly emphasized in this 2nd edition are the essential basics of electronics theory and practice. This is a guide that will give its reader the unique advantage of being able to keep up to date with the many rapid advances continuously taking place in the electronics field. It is an excellent reference for the beginner, student, or hobbyist. 240 pp., 173 illus.
**Paper $12.95   Hard $16.95
Book No. 2887**

Send $1 for the new TAB Catalog describing over 1300 titles currently in print and receive a coupon worth $1 off on your next purchase from TAB.

(In PA, NY, and ME add applicable sales tax. Orders subject to credit approval. Orders outside U.S. must be prepaid with international money orders in U.S. dollars.)

*Prices subject to change without notice.

To purchase these or any other books from TAB, visit your local bookstore, return this coupon, or call toll-free 1-800-233-1128 (In PA and AK call 1-717-794-2191).

Product No.	Hard or Paper	Title	Quantity	Price

☐ Check or money order enclosed made payable to TAB BOOKS Inc.

Charge my ☐ VISA ☐ MasterCard ☐ American Express

Acct. No. _____ Exp. _____

Signature _____

Please Print
Name _____

Company _____

Address _____

City _____

State _____ Zip _____

Subtotal	
Postage/Handling ($5.00 outside U.S.A. and Canada)	$2.50
In PA, NY, and ME add applicable sales tax	
TOTAL	

Mail coupon to:
TAB BOOKS Inc.
Blue Ridge Summit
PA 17294-0840

BC